Cobalt Compounds and Applications

Edited by Yasemin Yıldız and Aynur Manzak

Published in London, United Kingdom

IntechOpen

Supporting open minds since 2005

Cobalt Compounds and Applications
http://dx.doi.org/10.5772/intechopen.81986
Edited by Yasemin Yıldız and Aynur Manzak

Contributors
Alicia Boix, Leticia Gomez, Surender Kumar, Baljeet Singh, Riadh Marzouki, Mohsen Graia, Mohamed Faouzi Zid, Mahmoud A Sayed, Syam Sundar Lingala, Mirela Dragan, Stanica Enache, Mihai Varlam, Konstantin Petrov, Yasemin Yıldız

First published in London, United Kingdom, 2019 by IntechOpen
IntechOpen is the global imprint of INTECHOPEN LIMITED, registered in England and Wales, registration number: 11086078, 7th floor, 10 Lower Thames Street, London, EC3R 6AF, United Kingdom
Printed in Croatia

British Library Cataloguing-in-Publication Data
A catalogue record for this book is available from the British Library

Additional hard and PDF copies can be obtained from orders@intechopen.com

Cobalt Compounds and Applications
Edited by Yasemin Yıldız and Aynur Manzak
p. cm.
Print ISBN 978-1-78984-558-7
Online ISBN 978-1-78984-559-4
eBook (PDF) ISBN 978-1-83968-385-5

For EU product safety concerns:
IN TECH d.o.o., Prolaz Marije Krucifikse Kozulić 3, 51000 Rijeka, Croatia, info@intechopen.com or visit our website at intechopen.com.

Meet the editors

Yasemin Yıldız graduated from the Department of Chemistry at the Sakarya University. Yasemin Yildiz received her MSc degree from the Department of Chemistry at the Sakarya University in 2008. She worked as an analyst in research and development in the pharmaceutical industry. She started to work as a lecturer at Sakarya University in 2010. She received her PhD degree for Physical Chemistry from Sakarya University in 2014. Yasemin Yildiz is an active researcher in the field of supported and polymeric membranes, membrane stability, and membrane lifetime.

Aynur Manzak completed her BEng from the Faculty of Chemical and Metallurgical Engineering at Istanbul Technical University. After working in the perfume industry as a chemical engineer for about 4 years, she started her academic career as a research assistant in the Department of Chemistry at the Sakarya University. She received her MSc and PhD degrees from Sakarya University. Dr. Manzak worked as a visiting researcher at Clark University and Worcester Polytechnic Institute in Massachusetts, USA (2016-2017). She has been an Associate Professor in the Department of Chemistry at the Sakarya University since 2018. Her main research interests are separation science, membrane technology, emulsion liquid, and supported and polymer inclusion membrane.

Contents

Preface

The chemistry of cobalt compounds has attracted attention in recent years due to its applications in biological systems as antimicrobial agents and antibacterial agents. To design new drugs, medical chemistry can take advantage of the properties of cobalt ions and compounds. Cobalt studies continue because of their wide range of functions and applications in pharmaceutical technology. These studies have led to the emergence of medical bioorganometallic chemistry and the expansion of medical inorganic chemists. Studies in this field have led to new and difficult questions to date and this will continue. A recent literature review showed that many new cobalt compounds have been used in many different areas as scientists are seeking answers to these questions.

This book presents knowledge of the new cobalt compounds that can be used in many different applications for scientists, engineers, and students. The chapters include novel approaches of cobalt, nanostructured cobalt, new catalysts in some reactions, and biological systems and technologies.

We hope that the book will shed light on new horizons for scientists, engineers, and students.

Aynur Manzak
Department of Chemistry, Sakarya University,
Sakarya, Turkey

Yasemin Yildiz
Vocational School of Health Services, Sakarya University,
Sakarya, Turkey

Section 1

Introduction

Chapter 1

Introductory Chapter: Cobalt Compounds and Applications

Aynur Manzak and Yasemin Yildiz

1. Introduction

Cobalt has a subject of extensive research and application area due to being ferromagnetic with high thermostability, multivalent, a high-melting point (1493°C), and retaining its strength to a higher temperature [1]. Some of them are antimicrobial agents in biological systems, biocompatible magnetic-fluids, hybrid supercapacitors, magnetic resonance imaging and controlled drug delivery, nano-structured cobalt gas sensors, microwave absorbing paints and catalytic activity in some reactions, active sites for electrochemical applications [2].

Some of these extensive functions and applications can be mentioned. Cobalt ferrite ($CoFe_2O_4$) has a great physical and chemical stability and large anisotropy, making it suitable for biomedical applications. It has a magnetic property due to its tetrahedral (Td) and octahedral (Oh) sites, which include cation. Also, it is used to enhance the scope of materials in the biomedical field. Synthesis of $CoFe_2O_4$ is carried out by the following methods: sol-gel [3], solid-state reaction method [4], microemulsion [5], combustion [6], chemical coprecipitation [7], hydrothermal method [8], etc. [9].

The $CoFe_2O_4$ is a suitable material in energy harvesting/storage and conversion, pathogen detection, chemoresistive sensor, and dye degradation. Besides these features, it also has applications such as magnetic resonance imaging, magnetic fluid hyperthermia, drug delivery, and tissue repair. $CoFe_2O_4$ ferrite nanoparticles can be utilized in various antimicrobial applications that they have good antimicrobial activity against all tested bacteria, especially, Gram-negative bacteria. This feature of theirs can bring a new perspective to future studies in other ferrite and composite structures [10]. Also, the antimicrobial activity of $CoFe_2O_4$ was augmented with the addition of silver in some research [3, 11–13]. Kooti et al. [14] found that the antibacterial activity of Ag-coated $CoFe_2O_4$ nanocomposites is more efficient than Ag nanoparticles [9].

Cobalt oxide nanostructures have the matchless advantages of high theoretical capacity, highly active catalytic properties, and outstanding thermal/chemical stability. So, they are used as electrode materials for various electrochemical applications. Cobalt oxides are used in many research areas. They are abundant on earth and low cost and have environmental compatibility and excellent thermal stability and exceptional physical and chemical properties. Cobalt oxides have been successfully synthesized and used so far for various applications. These metal oxides stand out in applications of electrochemical energy devices. It will be used in theoretical calculations and simulations in the design of new cobalt oxide-based materials for practical applications in new studies. These studies encourage nanoscience and nanotechnology and the development of materials science, with regard to energy technologies [15].

In recent years, synthetic approaches have been developed for cobalt oxide-based nanomaterials [16]. Cobalt oxides are synthesized by a mild hydrothermal method using thioglycolic acid (TGA). Using hydrothermal method is provided with control of kinetic and thermodynamic factors [17]. At present, there is a quest to find the most suitable passivating material that can efficiently improve the performance of the cobalt oxide solar cell, and its success can be highly dependent on the understanding of the structures and properties of the passivating layer/cobalt oxide system. Further efforts should be made to clarify these mechanisms. Due to high crystallinity, size, morphology, optics, and good dispersion with Co_3O_4 particles, high solar cell yield was provided. The Co_3O_4 nanoparticles are preferred in low-temperature-fabricated solar cells for environmentally sensitive uses in the development of low-cost antireflection in processes [18].

In addition to all these studies, it is possible to see cobalt thin and ultrathin film structures in numerous new applications in industrial sectors. Cobalt has a wide range of uses due to its physical, mechanical, and electrical properties. Metallic cobalt films play an important role in the safety of integrated circuit devices. This is because metallic cobalt films have both a low tendency to diffusion exposure and have a higher resistance to electromigration.

The use of cobalt compounds in the biomedical field has recently been of interest to researchers. The use of cobalt/silica carriers to repair the retina removed in the relevant medical field is also under investigation [19].

In addition, one-dimensional layered $NH_4CoPO_4.H_2O$ microrods, which have high electrochemical activity, are low cost, and have natural richness, are obtained using a hydrothermal method. Although it utmost progress, it needs to studies for more stable and performance electrode materials for supercapacitors [20].

Then, microwave synthesis method was used to get bimetal phosphates, $NaNi_xCo_{1-x}PO_4.H_2O$; this method is favorable and timesaving in comparison with conventional hydrothermal [21], solvothermal [22], and pyrolysis methods [23]. Also, it satisfies a need for inexpensive electrodes in practical applications. This has proved that the combination of cobalt and nickel increases the specific capacity of the material. Cobalt can promote charge transfer owing to can act as electronic conductive species and can efficiently reduce the resistance and nickel can present high theoretical capacitance [21, 24–29]. At the end of the studies, optimized $NaNi_{0.33}Co_{0.67}PO_4.H_2O$ was obtained. The optimized $NaNi_{0.33}Co_{0.67}PO_4.H_2O$ exhibited a high electrochemical performance. The results show that it can be applied to practical devices that open a wide development potential for future supercapacitors.

In addition to all the mentioned properties of cobalt, it is a highly active catalyst for hydrogenation reactions. Supported in most cases, in low-temperature Fischer-Tropsch process, Co/Al_2O_3 or Co/SiO_2-ZrO_2 catalysts were used; CoO/SiO_2 and CoO/SiO_2-Al_2O_3 and CoO/ZrO_2-kieselguhr catalysts were used in the hydrogenation of oxoaldehydes, amination of alcohols, and amination of aldehydes and ketones for the production of ethylamine and propyl amines [30, 31].

It was determined that cobalt metal catalysts had higher activity when Ni and Co metal catalysts were compared. Cobalt-based catalysts exhibit high catalytic activity because the tendency of cobalt to react is higher due to its electron configuration. The results of the catalytic hydrolysis reaction of 2Co-1Ni-B/Magnesite catalyst and $NaBH_4$ synthesized by the co-loading of cobalt and nickel metals to the magnesite support material have increased the efficiency of the catalyst by adding Co [32].

Alloys are frequently used in the field of biomaterials of metal and metal alloys such as stainless steels, gold, and cobalt. Metallic biomaterials are preferred in the production of biomedical devices used for diagnostic and therapeutic purposes. The metals used in the prosthesis are suitable for the living body if used in small amounts [33]. Cobalt ferrite, barium, and strontium show promising magnetic behavior for medical applications [34]. Cobalt ferrite nanoparticles are more preferred for medical purposes than the other two nanoparticles. This is due to less toxic effects of cobalt ferrite. Cobalt ferrite has an application as a chip for diagnostic and laboratory applications and magnetic hyperthermia as minimally invasive tumor treatment [3, 35, 36].

Cobalt-chromium alloys are highly preferred materials in the biomedical field due to their high wear and corrosion resistance together with their load-bearing capabilities. Today, the surfaces of cobalt-chromium alloys are more biologically compatible with the surrounding tissue. That is, many studies have been done to make it bioactive. The surfaces of cobalt-chromium materials are tried to modify by being coated with bioactive glasses or metals with superior properties in terms of bioactivity. The cobalt-chromium alloy (ASTM F75) surfaces, which are used as orthopedic implant materials, were coated with titanium matrix composite material by cold dynamic gas spraying technique. Then, the thermal oxidation of the coatings was carried out in an air oven at 600°C for 60 hours [37].

All the features and applications we have mentioned so far prove how important cobalt is. The objective of this book is to serve as a one-stop comprehensive information resource on new cobalt and nanostructured cobalt compounds, from science to industry and health area.

Author details

Aynur Manzak[1] and Yasemin Yildiz[2*]

1 Department of Chemistry, Sakarya University, Sakarya, Turkey

2 Vocational School of Health Services, Sakarya University, Sakarya, Turkey

*Address all correspondence to: yyildiz@sakarya.edu.tr

References

[1] Darton Commodities Ltd. Cobalt Market Report [Internet]. July 2013 [Accessed: 06 July 2017]

[2] Chakradhary VK, Ansari A, Jaleel Akhtar M. Design, synthesis, and testing of high coercivity cobalt doped nickel ferrite nanoparticles for magnetic applications. Journal of Magnetism and Magnetic Materials. 2019;**469**:674-680

[3] Sanpo N, Berndt CC, Wen C, Wang J. Transition metal-substituted cobalt ferrite nanoparticles for biomedical applications. Acta Biomaterialia. 2013;**9**:5830-5837

[4] Rafferty A, Prescott T, Brabazon D. Sintering behaviour of cobalt ferrite ceramic. Ceramics International. 2008;**34**:15-21

[5] Choi EJ, Ahn Y, Kim S, An DH, Kang KU, Lee B-G, et al. Superparamagnetic relaxation in $CoFe_2O_4$ nanoparticles. Journal of Magnetism and Magnetic Materials. 2003;**262**:L198-L202

[6] Yan C-H, Xu Z-G, Cheng F-X, Wang Z-M, Sun L-D, Liao C-S, et al. Nanophased $CoFe_2O_4$ prepared by combustion method. Solid State Communications. 1999;**111**:287-291

[7] Pannaparayil T, Komarneni S. Synthesis and characterization of ultrafine cobalt ferrites. IEEE Transactions on Magnetics. 1989;**25**:4233-4235

[8] Daou TJ, Pourroy G, Bgin-Colin S, Grenche JM, Ulhaq-bouillet C, Legar P, et al. Hydrothermal synthesis of monodisperse magnetite nanoparticles. Chemistry of Materials. 2006;**18**:4399-4404

[9] Mahajan P, Sharma A, Kaur B, Goyal N, Gautam S. Green synthesized (*Ocimum sanctum* and *Allium sativum*) Ag-doped cobalt ferrite nanoparticles for antibacterial application. Vacuum. 2019;**161**:389-397

[10] Rashmi P. Sharma, Siddheshwar D. Raut, Ramjan M. Mulani, Ambadas S. Kadam, Rajaram S. Mane., Sol-gel auto-combustion mediated cobalt ferrite nanoparticles: A potential material for antimicrobial applications, International Nano Letters. 2019;**9**:141-147

[11] Velho-Pereira S, Noronha A, Mathias A, Zakane R, Naik V, Naik P, et al. Antibacterial action of doped $CoFe_2O_4$ nanocrystals on multidrug resistant bacterial strains. Materials Science and Engineering: C. 2015;**52**:282-287

[12] Xavier S, Cleetus H, Nimila P, Thankachan S, Sebastian R, Mohammed EM. Structural and antibacterial properties of silver substituted cobalt ferrite nanoparticles. Research Journal of Pharmaceutical, Biological and Chemical Sciences. 2014;**5**:364-371

[13] Shukla R, Ningthoujam R, Umare S, Sharma S, Kurian S, Vatsa R, et al. Decrease of superparamagnetic fraction at room temperature in ultrafine $CoFe_2O_4$ particles by Ag-doping. Hyperfine Interactions. 2008;**184**:217-225

[14] Kooti M, Saiahi S, Motamedi H. Fabrication of silver-coated cobalt ferrite nanocomposite and the study of its antibacterial activity. Journal of Magnetism and Magnetic Materials. 2013;**333**:138-143

[15] Mei J, Liao T, Ayoko GA, Bell J, Sun Z. Cobalt oxide-based nanoarchitectures for electrochemical energy applications. Progress in Materials Science. 2019;**103**:596-677

[16] Gogotsi Y, Penner RM. Energy storage in nanomaterials—Capacitive, pseudocapacitive, or battery-like? ACS Nano. 2018;**12**:2081-2083

[17] Sharma D, Rajput J, Kaith B, Kaur M, Sharma S. Synthesis of ZnO nanoparticles and study of their antibacterial and antifungal properties. Thin Solid Films. 2010;**519**:1224-1229. DOI: 10.1016/j.tsf.2010.08.073

[18] Ambika S, Gopinath S, Saravanan K, Sivakumar K, Ragupathi C, Sukantha TA. Structural, morphological and optical properties and solar cell applications of thioglycolic routed nano cobalt oxide material. Energy Reports. 2019;**5**:305-309

[19] Pankhurst QA, Connolly J, Jones SK, Dobson J. Applications of magnetic nanoparticles in biomedicine. Journal of Physics D: Applied Physics. 2003;**36**:R167-R181

[20] Liu M, Shang N, Zhang X, Gao S, Wang C, Wang Z. Microwave synthesis of sodium nickel-cobalt phosphates as high-performance electrode materials for supercapacitors. Journal of Alloys and Compounds. 2019;**791**:929-935

[21] Chang J, Xiao Y, Xiao M, Ge J, Liu C, Xing W. Surface oxidized cobalt-phosphide nanorods as an advanced oxygen evolution catalyst in alkaline solution. ACS Catalysis. 2015;**5**:6874-6878

[22] Liao M, Zeng G, Luo T, Jin Z, Wang Y, Kou X, et al. Three-dimensional coral-like cobalt selenide as an advanced electrocatalyst for highly efficient oxygen evolution reaction. Electrochimica Acta. 2016;**194**:59-66

[23] Ahn HS, Tilley TD. Electrocatalytic water oxidation at neutral pH by a nanostructured $Co(PO_3)_2$ anode. Advanced Functional Materials. 2013;**23**:227-233

[24] Liang B, Chen Y, He J, Chen C, Liu W, He Y, et al. Controllable fabrication and tuned electrochemical performance of potassium Co-Ni phosphate microplates as electrodes in supercapacitors. ACS Applied Materials & Interfaces. 2018;**10**:3506-3514

[25] Shu Y, Li B, Chen J, Xu Q, Pang H, Hu X. Facile synthesis of ultrathin nickel-cobalt phosphate 2D nanosheets with enhanced electrocatalytic activity for glucose oxidation. ACS Applied Materials & Interfaces. 2018;**10**:2360-2367

[26] Kim H, Park J, Park I, Jin K, Jerng SE, Kim SH, et al. Coordination tuning of cobalt phosphates towards efficient water oxidation catalyst. Nature Communications. 2015;**6**:8253-8264

[27] Chen G, Liaw SS, Li B, Xu Y, Dunwell M, Deng S, et al. Microwave-assisted synthesis of hybrid $Co_xNi_{1-x}(OH)_2$ nanosheets: Tuning the composition for high performance supercapacitor. Journal of Power Sources. 2014;**251**:338-343

[28] Qu L, Zhao Y, Khan AM, Han C, Hercule KM, Yan M, et al. Interwoven three-dimensional architecture of cobalt oxide nanobrush-graphene@ $Ni_xCo_{2x}(\text{-}OH)_{6x}$ for high-performance supercapacitors. Nano Letters. 2015;**15**:2037-2044

[29] Wang Z, Wu P, Lan L, Liu K, Hu Y, Ji S. Preparation, characterization and hydrodesulfurization performances of $Co\text{-}Ni_2P/SBA\text{-}15$ catalysts. Journal of Energy Chemistry. 2015;**24**:185-192

[30] Busca G. Heterogeneous Catalytic Materials. Amsterdam: Elsevier; 2014

[31] Chen B, Dingerdissen U, Krauter J, Rotgerink HL, Möbus K, Ostgard D, et al. New developments in hydrogenation catalysis particularly in synthesis of fine and intermediate

chemicals. Applied Catalysis A: General. 2005;**280**(1):17-46

[32] Seda Hoşgün, Kobalt ve nikel içerikli destekli katalizörlerin sentezi, karakterizasyonu ve sodyum borhidrürden hidroliz tepkimesi ile hidrojen üretiminde kullanılması, Kimya Mühendisliği Anabilim Dalı, Eskişehir Osmangazi Üniversitesi, Fen Bilimleri Enstitüsü; Aralık. 2018

[33] Pasinli A. Biyomedikal Uygulamalarda Kullanılan Biyomalzemeler. Makine Teknolojileri Elektronik Dergisi. 2004;(4):25-34

[34] Biehl P, von der Lühe M, Dutz S, Schacher FH. Synthesis, characterization, and applications of magnetic nanoparticles featuring polyzwitterionic coatings. Polymers. 2018;**10**:91. DOI: 10.3390/polym10010091

[35] Peeples B, Goornavar V, Peeples C, Spence D, Parker V, Bell C, et al. Structural, stability, magnetic, and toxicity studies of nanocrystalline iron oxide and cobalt ferrites for biomedical applications. Journal of Nanoparticle Research. 2014;**16**:2290

[36] Salunkhe AB, Khot VM, Thorat ND, Phadatare MR, Sathish CI, Dhawale DS, et al. Polyvinyl alcohol functionalized cobalt ferrite nanoparticles for biomedical applications. Applied Surface Science. 2013;**264**:598-604

[37] Doğukan Çetiner. Biyomedikal uygulamalar için ASTM F75 kobalt-krom alaşımının yüzey modifikasyonu, Malzeme Bilimi ve Mühendisliği Anabilim Dalı, İstanbul Teknik Üniversitesi, Fen Bilimleri Enstitüsü; Ocak 2015

Section 2

Cobalt Phosphates and Applications

Chapter 2

Cobalt Phosphates and Applications

Riadh Marzouki, Mahmoud A. Sayed, Mohsen Graia and Mohamed Faouzi Zid

Abstract

Cobalt phosphates with open framework present various physical performances in relation to their structures. In fact, the development of new materials that could potentially be ionic conductors or ion exchangers led us to examine the Co-P-O and A-Co-P-O crystallographic systems (A: monovalent cation) and their different methods of synthesis. This work consists first of all in highlighting the crystalline phases of cobalt phosphates. Indeed, many works related to the discovery of some of these materials with interesting properties, in particular ionic conductivity, motivated our research and encouraged us to collect several cobalt phosphates and to correlate structure-physical properties in particular electrical properties.

Keywords: cobalt phosphate, structure, open framework, physical property, battery

1. Introduction

The search for new inorganic materials with open frameworks formed by tetrahedra and octahedra sharing corners or edges; delimiting cages (1D), interlayer spaces (2D), or tunnels (3D); or communicating by the intermediate of windows where cations are located is an interesting field with intense activity including several disciplines: solid-state chemistry, physics, mechanics, etc. Synthesis and physicochemical studies of metallophosphate compounds are the driving force behind the recent technological development, and studies are progressing through the exchange of points of view between specialists concerned.

Metallophosphates have a promising field for various applications: electrical, electrochemical, magnetic, and catalytic processes [1–7]. Nevertheless, the introduction of monovalent ions into metallophosphates can lead to materials with interesting properties. This orientation was initiated from the discovery of the ionic conduction properties of NASICON $Na_3Zr_2Si_2PO_{12}$ ($\sigma_{300°C}$ = 0.2 S cm^{-1} and Ea = 0.29 eV) in 1976 [5] followed by olivine series studies of general formula $LiMPO_4$ (M = Co^{2+}, Fe^{2+}, Mn^{2+}) usable in the manufacture of cathodes of rechargeable lithium-ion batteries [7]. These materials have a remarkable structural richness: olivine structure [7], zeolitic structure [8], alluaudite structure [9], melilite structure [10], etc. In relation to their structures, these materials have many physicochemical properties: ionic conduction [10], ion exchange [6, 7], etc. In this context, several researcher groups have tried to explore CoO-P_2O_5 and A_2O-CoO-P_2O_5 systems (A: monovalent metals). This chapter is dedicated to treated physicochemical and structural studies of monovalent cation cobalt phosphates (Li, Na, K, and Ag).

2. Synthesis methods and experimental techniques

The most common synthesis method is the solid-state reaction method. Nevertheless, to minimize the energy consumption and to improve quality of the developed materials (particle size, purity, homogeneity, etc.), other techniques such as hydrothermal method are adopted. In this method, the crystalline products are synthesized at low temperature, generally 150–250°C, and under high pressure.

2.1 Solid-state reaction method

Solid-state reaction route is the most adopted method to prepare single crystals or polycrystalline materials. The essential steps are:

- Mixing and grinding solid reagents and placing the mixture in a container (usually porcelain, alumina, or platinum crucibles).

- Calcination: a first heat treatment at 573–673 K for a few hours to remove the volatile compounds (NH_3, H_2O, CO_2, etc.).

- Grinding another time the remaining mixture to homogenize and reduce the size of the particles which will increase the contact area between the grains.

- Second heat treatment by gradually increasing the temperature to a so-called "pasty" state of the mixture (partially melted mixture). Maintain this temperature for a few days, and then slowly lower it to room temperature.

2.2 Hydrothermal route

The hydrothermal or solvothermal method consists of preparing an aqueous solution containing the reagents dissolved totally or partially. The aqueous solution is transferred either into a Teflon autoclave, both enclosed in metal autoclave.

The preparation in the autoclave is brought to a temperature between 373 and 573 K maintained for a few days in order to obtain single crystals. The maximum temperature is imposed by the resistance of the material constituting the Teflon.

Note: In this chapter, structures have been determined using X-ray diffraction (on single crystal or on powder). Electrical measurements are carried out using often complex impedance spectroscopy.

2.3 Experimental techniques

In this chapter, the structural studies of the studied materials were carried out by X-ray diffraction on single crystals or in some cases X-ray powder diffraction.

Electrical measurements are often performed using the complex impedance spectroscopy technique.

3. Cobalt phosphate

In the literature, there are more than 80 allotropic forms of cobalt phosphates in which cobalt takes different oxidation degrees, sometimes in the same compound. Some cobalt phosphates have distinguishable physical properties in relation to their structures. In this chapter, cobalt monophosphate $CoPO_4$ will be reported.

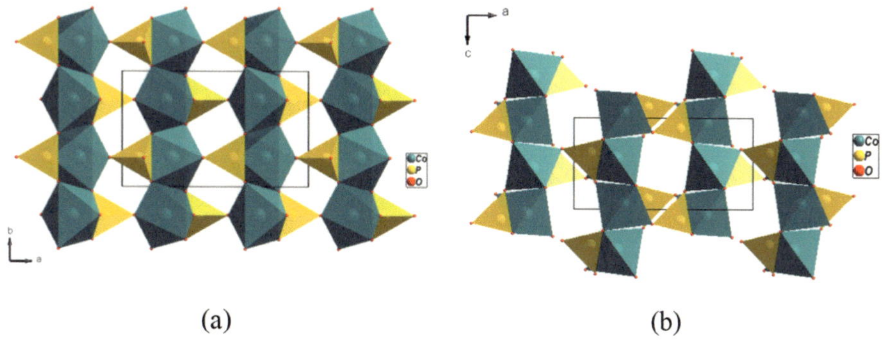

Figure 1.
Projection of $CoPO_4$ structure along the (a) c axis and (b) b axis.

$CoPO_4$ [1] material, like $FePO_4$ structure, is usable in the manufacture of Li-ion batteries. In fact, the lithium extraction from $LiCoPO_4$ material leads to $CoPO_4$ compound. The delithiated sample was prepared by electrochemical Li extraction in galvanostatic mode at a C/5-rate from $LiCoPO_4$. The latter shows considerable stability during several cycles of charge-discharge of the battery. In fact, $CoPO_4$ crystallized in the orthorhombic with Pnma space group. The structure is formed by $(CoO_6)_n$ chains connected with PO_4 tetrahedra to form layers in the *ab* plane. The connection between layers formed a 3D framework showing several types of tunnels according to [001] and [010] directions (**Figure 1**). In this structure type, the cobalt ion has an oxidation degree of +III.

4. Monovalent metal cobalt phosphate

There are more than 40 monovalent cation cobalt phosphates. The monovalent metal cobalt phosphates will be classified according to the oxygen/phosphor molar ratio.

4.1 Monophosphates

This family is known as orthophosphate or also monophosphate; it is characterized by its high stability compared to other phosphates. In the structure, $(PO_4)^{3-}$ tetrahedra are isolated from each other.

The most famous material is lithium cobalt monophosphate $LiCoPO_4$ (**Figure 2**) [1]. It crystallizes in the orthorhombic system, Pnma space group. It belongs to the olivine family of general formula $LiMPO_4$ (M = Fe, Ni, Co, and Mn). Xiang Huang et al. [11] have proposed hydrothermal synthesis method of this monophosphate which shows performance in terms of reaction yield and product homogeneity versus dry route. The phospho-olivine series is used in the manufacture of cathodes in Li-ion batteries [12]. $LiCoPO_4$-$CoPO_4$ system shows high stability during several charge-discharge cycles of the battery at room temperature (**Figure 3**). The olivine structure can be described as a compact hexagonal stack of *A-B-A-B-A*-type oxygen layers. The A = Na or Co cations occupy half of the octahedral sites AO_6 and the B = P cations 1/8 of the available tetrahedral P sites of PO_4 tetrahedra.

On the other hand, when lithium is substituted by sodium in different synthesis conditions, the monophosphate $NaCoPO_4$ may present in four allotropic forms [13, 14]. **Figure 4** groups the polymorphisms in sodium cobalt monophosphate. All sodium materials show open anionic frameworks containing tunnels which contain sodium cations. On the other hand, the structure of the $P2_1/n$ form where

Figure 2.
Projection of $LiCoPO_4$ structure along [010] direction.

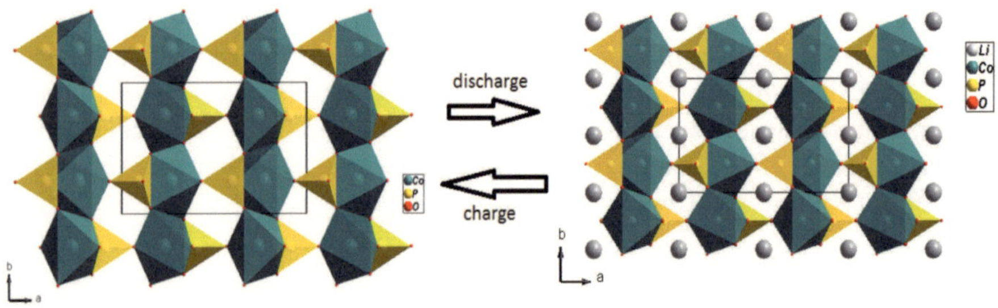

Figure 3.
Lithium insertion/extraction in the olivine structure $CoPO_4/LiCoPO_4$.

cobalt is only tetracoordinated is related to zeolite ABW ($LiAlSiO_4.H_2O$) [14]. In $NaCoPO_4$ ($P2_1/c$ space group subgroup of Pnma), Stucky et al. [13] report that the structure is also a distortion of the ABW zeolite structure but that it is a little more complex since the cobalt environment is trigonal bipyramidal. Indeed, the main characteristic of ABW zeolites is their spatial structures which contain pores and channels that can absorb or reject various solids, liquids, or gases. The applications of zeolites are numerous: food supplement for animals, additives for detergents, molecular filters, water treatment, catalysis, etc.

The α-$NaCoPO_4$ ($P2_1/n$ space group) with maricite type is formed by octahedral chains CoO_6 sharing edge and parallel to the *a* axis. They are interconnected via the PO_4 tetrahedra, which creates large cavities where Na^+ cations are located [13].

While the phase of the hexagonal system β-$NaCoPO_4$ is stuffed tridymite type which is a high temperature variety of quartz SiO_2. These compounds have a lower symmetry than tridymite due to the order of cations within the channels.

The silver cobalt monophosphate $AgCoPO_4$ [15] has another structure type with a twofold oxygen coordination for silver atoms and a fivefold coordination for cobalt atoms. Indeed, the silver compound crystallizes in the triclinic system, space group P-1. A projection of the structure of this phase is shown in **Figure 5**.

Another monophosphate is classified as Na-ionic conductor: $NaCo_4(PO_4)_3$ [16] with activation energy Ea = 0.89 eV and $\sigma = 10^{-6}$ S cm^{-1}. Indeed, cationic sites, located in wide-sectioned channels (**Figure 7a**), are partially occupied by Na^+ ions and relatively agitated which may explain the sodium mobility in the anionic framework. This compound crystallizes in the monoclinic system, space group $P2_1/n$. The

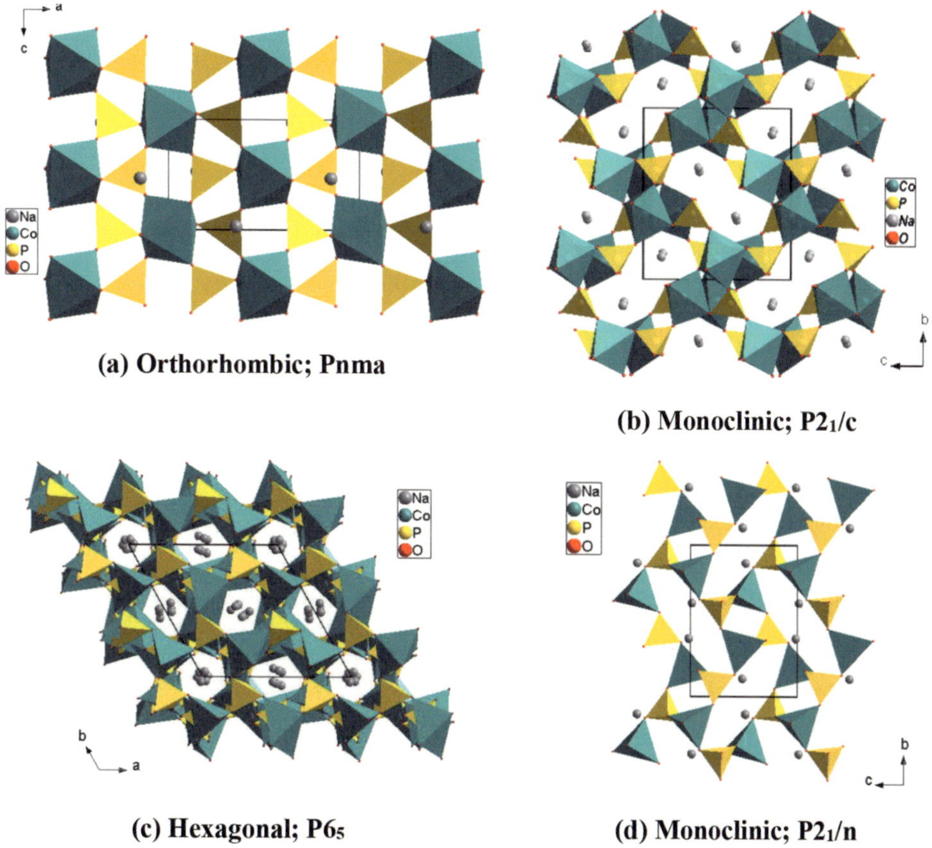

Figure 4.
Allotropic forms of $NaCoPO_4$: (a) Pnma, (b) $P2_1/c$, (c) $P6_5$, and (d) $P2_1/n$ space groups.

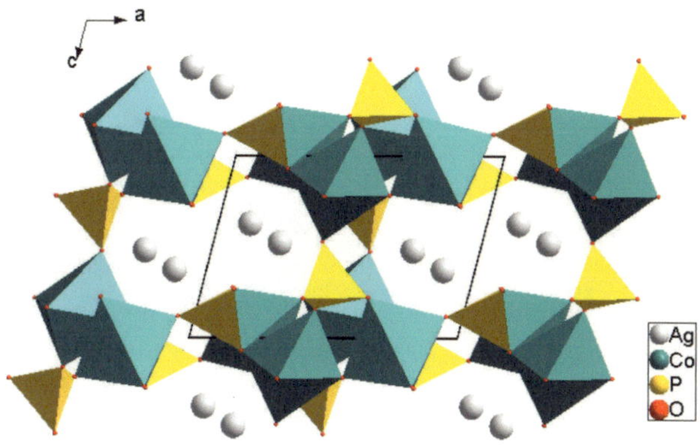

Figure 5.
Projection of $AgCoPO_4$ structure along [010] direction.

isoformula potassium material $KCo_4(PO_4)_3$ [17] crystallizes, in a different structure, the orthorhombic system, space group Pnnm. The structure projection along the [001] direction is shown in **Figure 6(b)**.

The sodium cobalt monophosphate $Na_4Co_7(PO_4)_6$ [18] is synthesized by the dry route. This compound is a member of a family of phases including $Na_4Ni_7(PO_4)_6$ [19] and $K_4Ni_7(PO_4)_6$ [20]. Previous studies have shown that the material $Na_4Ni_7(PO_4)_6$ is classified as fast ionic conductor. Several studies relating to the

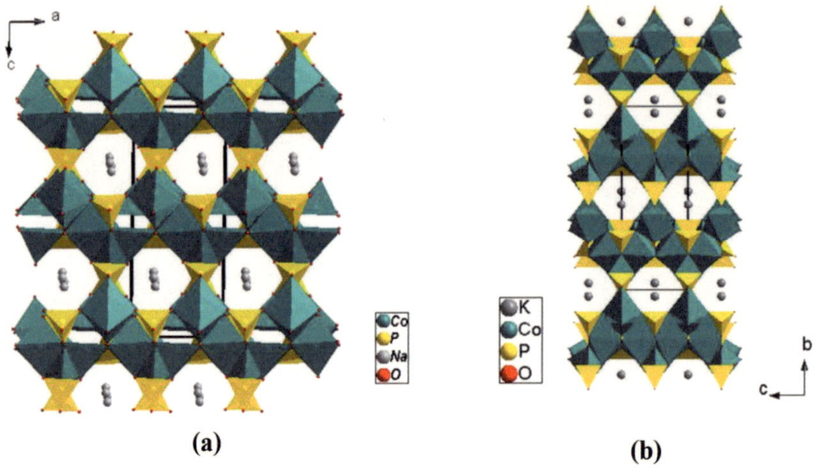

Figure 6.
Projections of (a) $NaCo_4(PO_4)_3$ and (b) $KCo_4(PO_4)_3$ structures.

substitution of phosphate by arsenate have led to $Na_4Co_7(AsO_4)_6$ (Ea = 1.00 eV) [21], $Na_4Co_{5.63}Al_{0.96}(AsO_4)_6$ (Ea = 0.53 eV) [22–24], $Na_4Li_{0.62}Co_{5.67}Al_{0.71}(AsO_4)_6$ [25], and $Ag_4Co_7(AsO_4)_6$ (Ea =0.61 eV) [26].

A projection of the structure of $Na_4Co_7(PO_4)_6$ according to [100] is given in **Figure 7**. The anionic framework has both a tetrahedral (CoO_4 and PO_4) and octahedral (CoO_6) environment as well as hexagonal tunnels where the sodium ions lodge.

4.2 Polyphosphates

Short-chain polyphosphates also named n-polyphosphates are characterized by short chains of PO_4^{3-} tetrahedra sharing corners. The general formulas of the phosphate anion are given by $[P_nO_{3n+1}]^{(n+2)-}$ with n > 1. Oligophosphates for which n = 2, 3, 4, and 5 are known until now. These compounds are infrequent for n ≥ 4.

Figure 7.
Projection of $Na_4Co_7(PO_4)_6$ structure along the a axis.

The other type corresponds to polyphosphates with long chains. When n tends to infinity, their phosphate anions take the formula $[PO_3]_n^{n-}$, thus forming infinite chains of PO_4 tetrahedra. If the tetrahedron chain closes on itself to form rings, the corresponding phosphates are called cyclophosphates. The general formula of the cyclic anion is $[P_nO_{3n}]^{n-}$ with n = 3, 4, 5, 6, 8, 10, and 12.

In this part, a variety of monovalent ion cobalt polyphosphates found in the literature will be mentioned.

4.2.1 Short-chain polyphosphates $[P_nO_{3n+1}]^{(n+2)-}$ with n > 1

4.2.1.1 Diphosphates (n = 2)

The formula of the phosphate anion is $P_2O_7^{4-}$, known as diphosphate (or pyrophosphate). The group P_2O_7 consists of two PO_4 tetrahedra sharing a single corner.

Concerning diphosphates of formulation $A_2CoP_2O_7$ (A: Na or K), the sodium compound is presented in three allotropic forms: triclinic (**Figure 8a**), monoclinic (**Figure 8b**), and quadratic (**Figure 8c**) [8, 10].

In the last form, cobalt atoms have purely tetrahedral environment, and the anionic framework is formed by layers formed by $[CoP_2O_7]^{2-}$ groups. The Na^+ and/or K^+ cations are located in the interlayer space. Sanz et al. [10] postponed the study of the ionic conductivity of the quadratic form to sodium; their study reveals that it is a fast ionic conductor. Marzouki et al. [27] proposed a modeling of

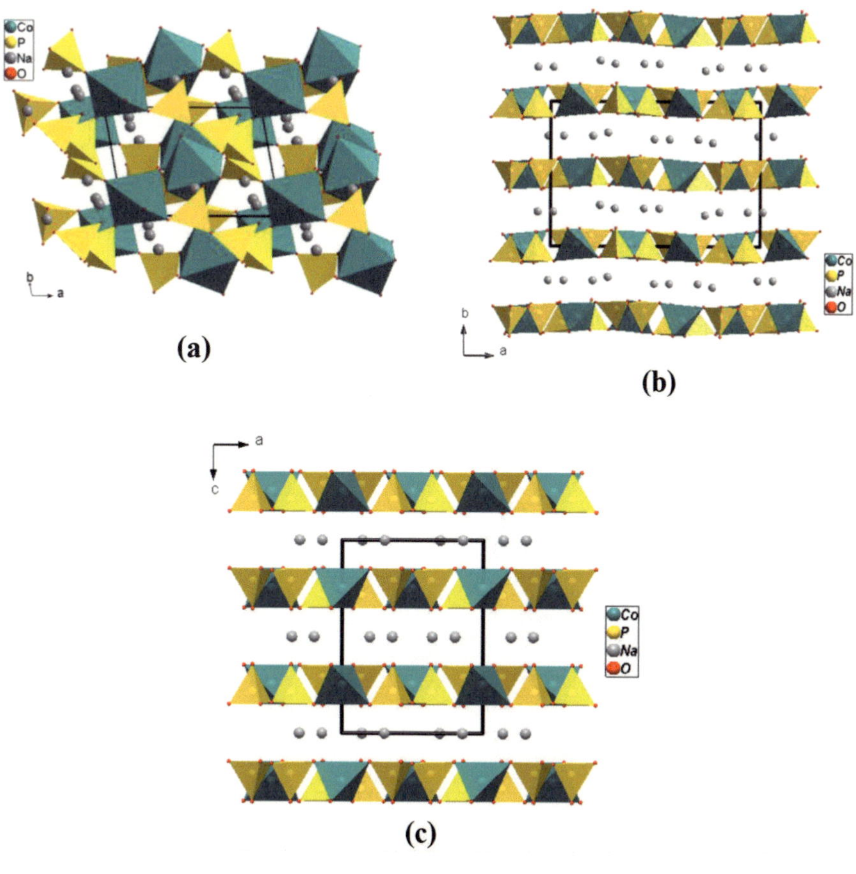

Figure 8.
Projections of polymorphs of $Na_2CoP_2O_7$: (a) triclinic, (b) monoclinic, and (c) tetragonal.

alkaline cation conduction paths in these structures (**Figure 9**). The conductivity in this type of material is bi-dimensional.

The silver cobalt diphosphates include $Ag_{3.68}Co_2(P_2O_7)_2$ [28] and $(Ag_{0.58}Na_{1.42})_2Co_2(P_2O_7)_2$ [29]. They crystallize in the triclinic system, space group P-1. Projection of the mixed Na/Ag metals is presented in **Figure 10**. The cobalt, in this case, is purely octahedral. In the anionic framework, the cohesion between two symmetrical units $Co(2)P_2O_{11}$ is provided by $Co(1)O_6$ octahedra to form the $Co_4P_4O_{28}$ unit. According to the three spatial directions, the junction between two $Co_4P_4O_{28}$ units is provided by two P_2O_7 diphosphates forming 3D anionic framework.

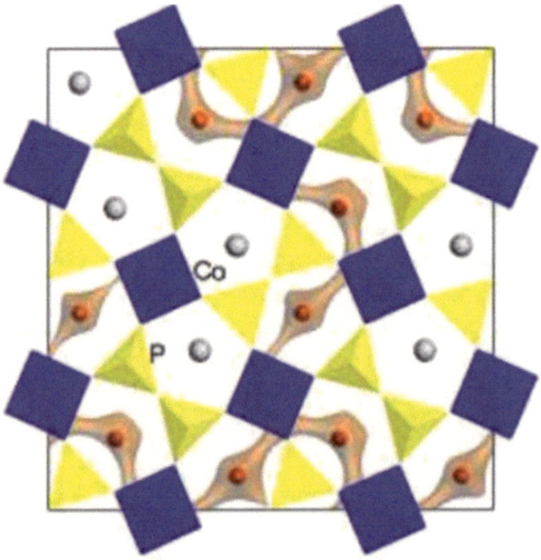

Figure 9.
Bond valence site energy-simulated pathways of Na^+ ions within the $K_{0.86}Na_{1.14}CoP_2O_7$ structure (Na brown and K gray and the layer at z = 0).

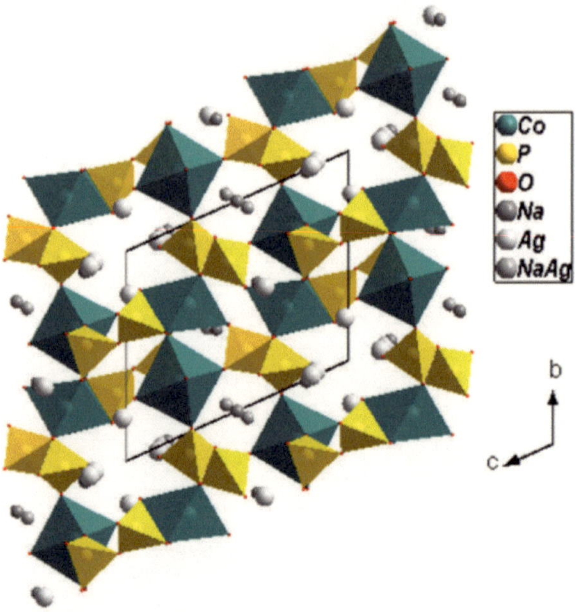

Figure 10.
Projection of $(Ag_{0.58}Na_{1.42})_2Co_2(P_2O_7)_2$ structure along the a axis.

Silver transport pathways in $Ag_{3.68}Co_2(P_2O_7)_2$ are simulated using BVSE calculations. The BVSE simulation shows that the material should be moderate 3D ionic conductor with activation energy value of 1.7 eV. The result is described in **Figure 11**.

The particularity of lithium cobalt diphosphates is the non-stoichiometry in composition. The formulas found in the bibliography are $Li_{5.88}Co_{5.06}(P_2O_7)_4$ [30] where cobalt and lithium share the same crystallographic sites and $Li_{4.03}Co_{1.97}(P_2O_7)_2$ [31] where a fraction of cobalt oxidation degree is +III. The projections of their structures are shown in **Figures 12** and **13**.

4.2.1.2 Triphosphates (n = 3)

Single monovalent cation cobalt triphosphate is found in the literature. Its formula is $LiCo_2P_3O_{10}$ [32]. This material crystallizes in the monoclinic system, space group $P2_1/m$. In the anionic framework, the P_3O_{10} groups ensure cohesion between the infinite chains formed by Co_2O_{10} dimers which are linked

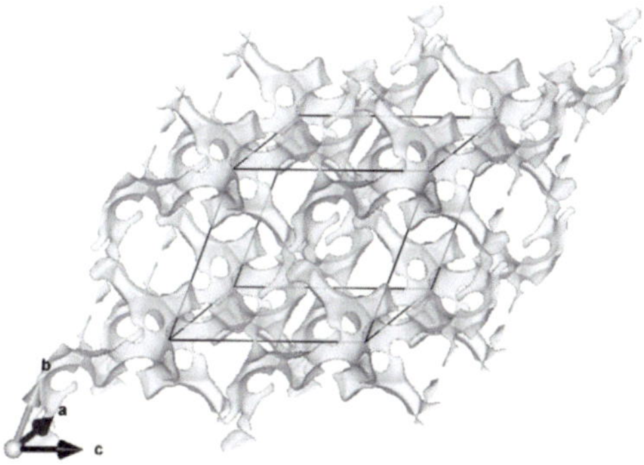

Figure 11.
3D silver transport pathways in $Ag_{3.68}Co_2(P_2O_7)_2$ with bond valence mismatch of $|\Delta V(Ag)| = 1.3$ u.v (i.e., ~1.7 eV).

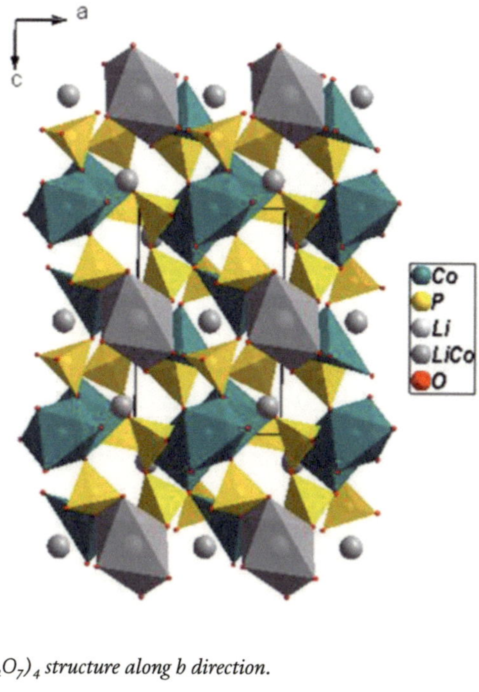

Figure 12.
Projection of $Li_{5.88}Co_{5.06}(P_2O_7)_4$ structure along b direction.

Figure 13.
Projection of $Li_{4.03}Co_{1.97}(P_2O_7)_2$ structure along a direction.

together by edge sharing. **Figure 14** shows a projection of the structure in the direction [100]. The $NaCo_2As_3O_{10}$ triarsenate [33], isostructural with $LiCo_2P_3O_{10}$ triphosphate, shows interesting electrical properties (Ea = 0.48 eV; $\sigma_{300°C} = 1.2 \times 10^{-5}$ S cm^{-1}).

4.2.1.3 Polyphosphates with long chains $[PO_3]_n^{n-}$

The first phase seen in the bibliography is tetraphosphate $K_2Co(PO_3)_4$ [34]. This material is synthesized by the dry route; it crystallizes in the monoclinic system, with non-centrosymmetric space group "Cc." Cobalt has the oxidation state (+II) and is octacoordinated. Phosphate anions of formulation $[PO_3]_n^{n-}$ (n tends to

Figure 14.
Projection of $LiCo_2P_3O_{10}$ structure along a direction.

infinity) thus develop into long chains of PO_4 tetrahedra linked together by CoO_6 octahedra to form a 3D framework (**Figure 15**).

As for the lithium compound [35] with $LiCo_2P_3O_9$ formula (**Figure 16**), this material is at a higher symmetry: orthorhombic system, space group P212121. The Co_2O_{11} dimers, in this case, are formed by two vertex-linked CoO_6 octahedra. They ensure the cohesion between the nn-infinite tetrahedral (PO_3) chains to lead to a three-dimensional framework.

4.2.1.4 Mixed mono-diphosphate $Na_4Co_3(PO_4)_2P_2O_7$

Some materials may have more than one type of phosphate group. The material $Na_4Co_3(PO_4)_2P_2O_7$ [36] has both PO_4 isolated tetrahedra and P_2O_7 diphosphate groups. This mono-diphosphate crystallizes in the orthorhombic system, space

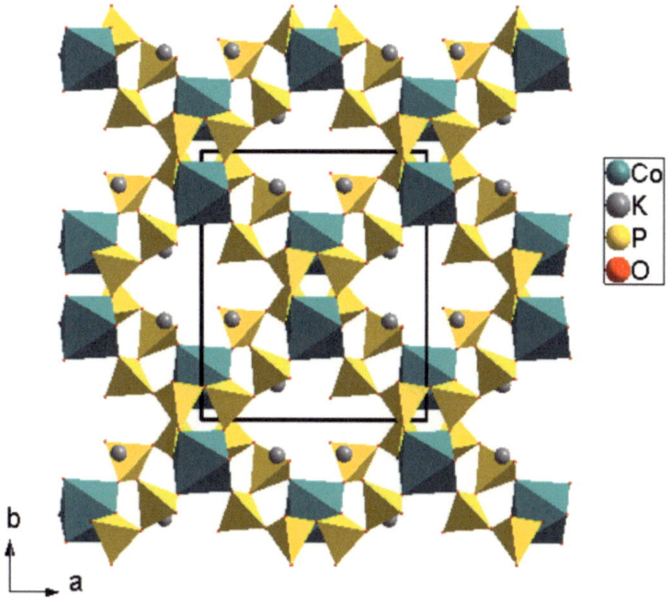

Figure 15.
Projection of $K_2Co(PO_3)_4$ structure along the c axis.

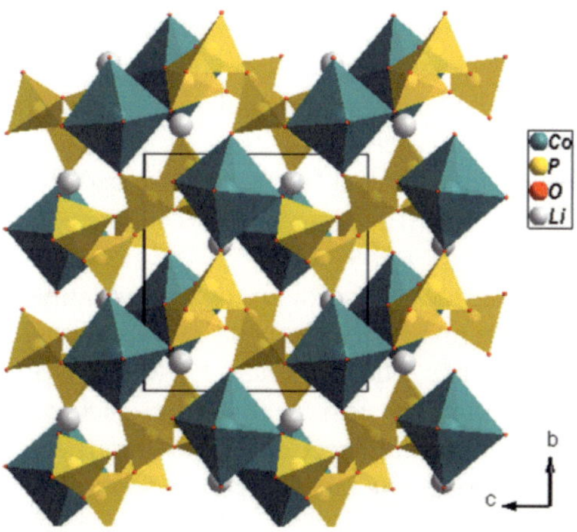

Figure 16.
Projection of $LiCo_2P_3O_9$ structure along the a axis.

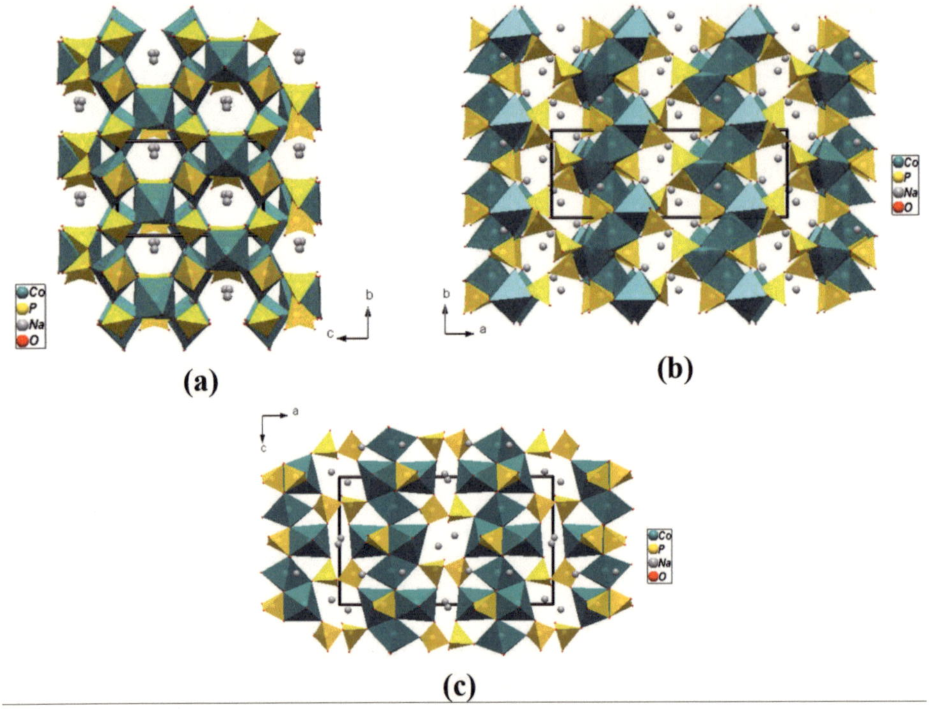

Figure 17.
Projections of $Na_4Co_3(PO_4)_2P_2O_7$ structure in (a) a, (b) b, and (c) c directions.

group $Pn2_1a$. Projections of the three-dimensional framework of this material (**Figure 17**) show that PO_4 monophosphates are bound to CoO_6 octahedra, on the one hand by edge sharing and on the other hand by sharing vertices, while diphosphates join four CoO_{10} units by pooling vertices.

5. Structural factors involved in ionic conductivity: Correlation of structure-electrical properties

Ionic conductors have been intensively researched since the discovery of properties of ionic superconductors [37] whose conductivity is sufficiently high to consider applications as solid electrolytes in batteries [38, 39] in storage devices of energy and sensors [40]. On the other hand, materials with low ionic conductivity remain interesting to elucidate certain mechanisms of cation transport. In these materials, the charge carriers are cations.

The open framework is an essential factor that governs the mobility of cations within a crystal lattice [6, 41, 42]. Among these structures, there are:

- Three-dimensional frameworks with windows or channels: this type of material has an ionic conduction influenced by the size of the bottlenecks separating two adjacent available sites. The existence of wide-sectioned channels between the cationic sites promotes the passage of cations. According to Hong, for fast ionic conduction, the minimum sections of the windows must be greater than or equal to twice the sum of the radii of the cation and the nearest anion [40].

- Layered structures: in this case, mobile ions move in parallel planes, located in the interlayer space. The conduction in this case is probably two-dimensional [27].

- Structures with isolated tetrahedral groups: these structures consist of tetrahedral groups (SiO_4^{3-}, PO_4^{3-}, etc.) connected to each other solely by alkaline ions. These independent tetrahedra facilitate the movement of cations [43–45].

Other factors than the open framework can also promote ionic mobility [40]:

- Site occupation: the partial occupations of the ionic sites (occupancy rate lower than 1) favor the displacement of the mobile ion from one site to another energetically equivalent.

- Coordination polyhedra: the cation environment can play an important role in its mobility. Indeed, mobile ions can cross rectangular faces more easily than triangular faces.

- Ion size: to promote conduction, congestion must be minimized, so the use of small cations is recommended, to facilitate their movement.

- Structural defects: substitution or doping of one or more elements with other(s) having different degrees of oxidation is responsible for the creation of cationic vacancies at the origin of conduction properties in certain materials.

Taking into account the structural factors influencing the conductivity mentioned above, several studies have been devoted to improving the electrical properties of such materials by acting on other factors. This is the case of the total or partial substitution of the mobile species such as in NASICON $Na_{1+x}Zr_{2-x}Mg_{x/2}(PO_4)_3$ ($0 < x \leq 2$) [46] and in SKELETON phosphates (3D) $A_3M_2(PO_4)_3$: A = Li, Na, Ag, K, and M = Cr, Fe [47]. The doping of materials by one or more chemical elements can also promote the mobility of cations like the oxides $La_{1.2}Sr_{1.8}Mn_{2-x}T_xO_7$ with T = Fe, Co, Cr [48].

On the other hand, work on a series of materials is being processed in order to show the effect of microstructure optimization (grain size) on conductivity [49]. Moreover, it has been demonstrated in previous studies, such as for LAMOX ceramics ($La_{2-x}R_xMo_{2-y}W_yO_9$ with R = Nd, Gd, Y) [50, 51] and for β-Xenophyllite-type $Na_4Co_7(AsO_4)_6$ [21] and $Ag_4Co_7(AsO_4)_6$ [26], that the electrical properties are related to the relative density of sample (100 porosity), which requires a rigorous control of the microstructure.

6. Conclusion

In this chapter, synthesis methods of cobalt phosphates and metallo-cobalt phosphates in the crystalline form have been described: single crystals and/or polycrystalline powders. The structural studies of the studied compounds show structural diversity with open anionic frameworks showing tunnels (3D) and inter-sheet space (2D). However, it shows that the electrical property is related to the structural characteristics of the material. In order to correlate structure and physical properties especially electrical properties of metallo-cobalt phosphates, structural factors influencing the ionic conductivity have been treated. Based on the structural characteristics, the electrical properties of the crystalline materials can be modeled theoretically, especially in the case of purely ionic conductors. In fact, it is possible to determine the value of the activation energy which corresponds to the minimum energy that must be supplied to an ion to move from one site to another site in the crystal lattice. In addition, these modelizations are based on the structural data of the crystal. Since the measurements are often performed on ceramics, it is also necessary to take into account the effect of the relative density of the ceramic: effect of the microstructure.

Author details

Riadh Marzouki[1,2,3*], Mahmoud A. Sayed[4,5], Mohsen Graia[2] and Mohamed Faouzi Zid[3]

1 Chemistry Department, College of Science, King Khalid University, KSA

2 Chemistry Department, Faculty of Sciences, University of Sfax, Tunisia

3 Chemistry Department, Faculty of Sciences, University of Tunis El Manar, Tunisia

4 Physics Department, Faculty of Science, King Khalid University, KSA

5 Physics Department, Faculty of Science, Al-Azher University, Assiut, Egypt

*Address all correspondence to: riadh.marzouki@hotmail.fr

References

[1] Ehrenberg H, Bramnik NN, Senyshin A, Fuess H. Crystal and magnetic structures of electrochemically delithiated $Li_{1x}CoPO_4$ phases. Solid State Sciences. 2009;**11**:18-23. DOI: 10.1016/j.solidstatesciences.2008.04.017

[2] Nagpure M, Shinde KN, Kumar V, Ntwaeaborwa OM, Dhoble SJ, Swart HC. Combustion synthesis and luminescence investigation of $Na_3Al_2(PO_4)_3$:RE (RE = Ce^{3+}, Eu^{3+} and Mn^{2+}) phosphor. Journal of Alloys and Compounds. 2010;**492**:384-388. DOI: 10.1016/j.jallcom.2009.11.110

[3] Chen JG, Ang L, Wang C, Wei Y. Sol-gel preparation and electrochemical properties of $Na_3V_2(PO_4)_2F_3$ composite cathode material for lithium ion batteries. Journal of Alloys and Compounds. 2009;**478**:604-607. DOI: 10.1016/j.jallcom.2008.11.147

[4] Kanazawa T. Inorganic Phosphate Materials, Materials Science Monographs. Oxford, United Kingdom: Elsevier Science & Technology; 1989

[5] Hong SYP. Crystal structures and crystal chemistry in the system $Na_{1+x}Zr_2Si_xP_{3-x}O_{12}$. Materials Research Bulletin. 1976;**11**:173-182. DOI: 10.1016/0025-5408(76)90073-8

[6] Goodenough JB, Hong HY-P, Kafalas JA. Fast Na^+-ion transport in skeleton structures. Materials Research Bulletin. 1976;**11**:203-220. DOI: 10.1016/0025-5408(76)90077-5

[7] Padhi AK, Nanjundaswamy K, Goodenough JB. Phospho-olivines as positive-electrode materials for rechargeable lithium batteries. Journal of The Electrochemical Society. 1997;**144**(4):1188-1194. DOI: 10.1149/1.1837571

[8] Erragh F, Boukhari A, Elouadi B, Holt EM. Crystal structures of two allotropic forms of $Na_2CoP_2O_7$. Journal of Crystallographic and Spectroscopic Research. 1991;**21**:321-326. DOI: 10.1007/BF01156084

[9] Ranko PB, Slavi CS. Hydrothermal synthesis and characterization of two ammonium-templated cobalt aluminophosphates. Journal of Materials Chemistry. 1999;**9**:2679-2682. DOI: 10.1039/A904587B

[10] Sanz F, Parada C, Rojo JM, Ruiz-Valero C, Saez-Puche R. Studies on tetragonal $Na_2CoP_2O_7$, a novel ionic conductor. Journal of Solid State Chemistry. 1999;**145**:604-611. DOI: 10.1006/jssc.1999.8249

[11] Xiang H, Junfeng M, Pingwei W, Yingmo H, Jinhui D, Zhibin Z, et al. Hydrothermal synthesis of $LiCoPO_4$ cathode materials for rechargeable lithium ion batteries. Materials Letters. 2005;**59**:578-582. DOI: 10.1016/j.matlet.2004.10.049

[12] Shigeto O, Shoichiro O, Minato E, Jun-ichi Y, Mitsuharu T, Hiroyuki K, et al. Cathode properties of phospho-olivine $LiMPO_4$ for lithium secondary batteries. Journal of Power Sources. 2001;**97-98**:430-432. DOI: 10.1016/S0378-7753(01)00631-0

[13] Feng P, Bu Xian-Hui GD, Stucky GD. Synthesis, crystal structure, and magnetic properties of a new polymorphic sodium cobalt phosphate with trigonal bipyramidal Co^{2+} and a tunnel structure. Journal of Solid State Chemistry. 1997;**129**:328-333. DOI: 10.1006/jssc.1996.7260

[14] Chippindale AM, Cowley AR, Chen J, Gao Q, Xu R. A new modification of $NaCoPO_4$ with the zeolite ABW structure. Acta Crystallographica C. 1999;**55**:845-847. DOI: 10.1107/S010827019900414X/br1238sup2.hkl

[15] Tordjman I, Guitel JC, Durif A, Averbuch MT, Masse R. Structure cristalline du monophosphate $AgCoPO_4$. Materials Research Bulletin. 1978;**13**:983-988. DOI: 10.1016/0025-5408(78)90111-3

[16] Baies R, Perez O, Caignaert V, Pralong V, Raveau B. A new sodium cobaltophosphate with a tunnel structure, ionic conductor. Journal of Materials Chemistry. 2006;**16**: 2434-2438. DOI: 10.1039/B516383H

[17] Lopez ML, Daidouh A, Pico C, Carvajal JR, Veiga ML. A comparison of the magnetic structures of $KMn_4(PO_4)_3$ and $KCo_4(PO_4)_3$ based on the connectivity of the coordination polyhedral. Chemistry: A European Journal. 2008;**14**(34):10829-10838. DOI: 10.1002/chem.200800763

[18] Kobashi D, Kohara S, Yamakawa J, Kawahara A. Un monophosphate synthétique de sodium et de cobalt: $Na_4Co_7(PO_4)_6$. Acta Crystallographica C. 1998;**54**:7-9. DOI: 10.1107/S010827019701189X

[19] Moring J, Kostiner E. The crystal structure of $Na_4Ni_7(PO_4)_6$. Journal of Solid State Chemistry. 1986;**62**:105-111. DOI: 10.1016/0022-4596(86)90221-5

[20] Ben Smail R. Préparation, cristallochimie et conduction ionique de phosphates et arséniates doubles de nickel et d'ions alcalins ou pseudo-alcalins [Thèse de Doctorat en Chimie]. FST; 2003

[21] Ben Smida Y, Marzouki R, Georges S, Kutteh R, Avdeev M, Guessmi A, et al. Synthesis, crystal structure, electrical properties, and sodium transport pathways of the new arsenate $Na_4Co_7(AsO_4)_6$. Journal of Solid State Chemistry. 2016;**239**:8-16. DOI: 10.1016/j.jssc.2016.04.005

[22] Marzouki R, Guesmi A, Driss A. The novel arsenate $Na_4Co_{7-x}Al_{2x/3}(AsO_4)6$ (x = 1.37): Crystal structure, charge-distribution and bond-valence-sum investigation. Acta Crystallographica C. 2010;**66**:i95-i98. DOI: 10.1107/S0108270110037376

[23] Marzouki R, Guesmi A, Zid MF, Driss A. Etude phisico-chimiques du monoaséniate mixte $Na_4Co_{5.63}Al_{0.91}(AsO_4)_6$ et simulation des chemins de conduction. Annales de Chimie-Science des Matériaux. 2013;**38**(3-4):117-129

[24] Marzouki R, Guesmi A, Zid MF, Driss A. Electrical properties and Na+ mobility in $Na_4Co_{5.63}Al_{0.91}(AsO_4)_6$ material. Journal of Inorganic Chemistry. 2013;**1**(1):9-16

[25] Marzouki R, Frigui W, Guesmi A, Zid MF, Driss A. β-Xenophyllite-type $Na_4Li_{0.62}Co_{5.67}Al_{0.71}(AsO_4)_6$. Acta Crystallographica Section E: Crystallographic Communications. 2013;**69**:i65-i66. DOI: 10.1107/S1600536813025233

[26] Marzouki R, Guesmi A, Georges S, Zid MF, Driss A. Structure, sintering and electrical properties of new ionic conductor $Ag_4Co_7(AsO4)_6$. Journal of Alloys and Compounds. 2014;**585**:74-79. DOI: 10.1016/j.jallcom.2013.09.213

[27] Marzouki R, Ben Smida Y, Guesmi A, Georges S, Ali IH, Adams S, et al. Structural and electrical investigation of new melilite compound $K_{0.86}Na_{1.14}CoP_2O_7$. International Journal of Electrochemical Science. 2018;**13**:11642-11662. DOI: 10.20964/2018.12.75

[28] Ben Moussa MA, Marzouki R, Brahmia A, Georges S, Obbade S, Zid MF. Synthesis and structure of new mixed silver cobalt(II)/(III) diphosphate—$Ag_{3.68}Co_2(P_2O_7)_2$. Silver(I) transport in the crystal. International Journal of Electrochemical Science. 2019;**14**:1500-1515. DOI: 10.20964/2019.02.55

[29] Marzouki R, Guesmi A, Zid MF. Synthesis, crystal structure and electrical properties of a new mixed compound $(Na_{0.71}Ag_{0.29})_2CoP_2O_7$. Crystal Structure Theory and Applications. 2012;**1**:68-73. DOI: 10.4236/csta.2012.13013

[30] Sanz F, Parada C, Ruiz-Valero C. Crystal growth, crystal structure, and magnetic properties of a new lithium cobalt diphosphate. Chemistry of Materials. 2000;**12**:671-676. DOI: 10.1021/cm991085+

[31] Kouass S, Guesmi A, Driss A. Le phosphate de cobalt et de lithium à valence mixte $Li_{4+x}Co_{2-x}(P_2O_7)_2$ (x = 0.03): Étude structurale et analyse de distribution de charge. Acta Crystallographica C. 2010;**66**:i4-i6. DOI: 10.1107/S0108270109052810

[32] Erragh F, Boukhari A, Holt EM. Lithium dicobalt tripolyphosphate and lithium dinickel tripolyphosphate. Acta Crystallographica C. 1996;**52**: 1867-1869. DOI: 10.1107/S010827019501643X

[33] Ben Smida Y, Marzouki R, Guessmi A, Georges S, Zid MF. Synthesis, structural and electrical properties of a new cobalt arsenate $NaCo_2As_3O_{10}$. Journal of Solid State Chemistry. 2015;**221**:132-139. DOI: 10.1016/j.jssc.2014.09.029

[34] Laugt M, Tordjman M, Bassi M, Guitel MC. Affinement des structures cristallines de $CuK_2(PO_3)_4$ et $CoK_2(PO_3)_4$. Acta Crystallographica. Section B. 1974;**30**:1100-1104. DOI: 10.1107/S056774087400433X

[35] Seungdon C, Seung-Tae H. Synthesis, crystal structure and magnetic properties of a new lithium cobalt metaphosphate, $LiCo(PO_3)_3$. Materials Research Bulletin. 2005;**40**:1787-1795. DOI: 10.1016/j.materresbull.2005.05.021

[36] Sanz F, Parada C, Amador U, Monge MA, Ruiz-Valero C. $Na_4Co_3(PO_4)_2P_2O_7$, a new sodium cobalt phosphate containing a three-dimensional system of large intersecting tunnels. Journal of Solid State Chemistry. 1996;**123**:129-139

[37] Yao YFY, Kummer JT. Ion exchange properties of and rates of ionic diffusion in beta-alumina. Journal of Inorganic and Nuclear Chemistry. 1967;**29**:2453

[38] Bauerle J. Study of solid electrolyte polarization by a complex admittance method. Physics and Chemistry of Solids. 1961;**30**:2657-2670

[39] Ravine D, Souquet J. Crystal structures and crystal chemistry in the system $Na_{1+x}Zr_2Si_xP_{3-x}O_{12}$. Journal of Chemical Physics. 1974;**71**:693-701

[40] Hong H-YP. Crystal structures and crystal chemistry in the system $Na_{1+x}Zr_2Si_xP_{3-x}O_{12}$. Materials Research Bulletin. 1976;**11**:173-182

[41] Boilot J, Colomban P, Théry, G. Collin J, Comès R. Relation structure-conductivité ionique dans les composés de type alumine β. Journal de Physique Colloques. 1978;**39**(C2):C2-204-C2-213. ff10.1051/jphyscol:1978235ff. ffjpa-00217390f

[42] Ravain D. Laboratoire d'énergétique Electrochimique, ENS d'électrochimie de Grenoble, Séminaire de Chimie du Solide; 1979

[43] Hu YW, Raistrick ID, Huggins RA. Ionic conductivity of lithium phosphate-doped lithium orthosilicate. Materials Research Bulletin. 1976;**11**:1227-1230

[44] Johnson RT, Biefeld RM, Keck JD. Ionic conductivity in Li_5AlO_4 and LiOH. Materials Research Bulletin. 1977;**12**:577-587

[45] Liebert BE, Huggins RA. Ionic conductivity of Li_4GeO_4, Li_2GeO_3 and $Li_2Ge_7O_{15}$. Materials Research Bulletin. 1976;**11**:533-538

[46] Cherkaoui F, Viala JC, Delmas C, Hagenmuller P. Crystal chemistry and ionic conductivity of a new Nasicon-related solid solution $Na_{1+x}Zr_{2-x}Mg_{x/2}(PO_4)_3$. Solid State Ionics. 1986;**21**:333-337. DOI: 10.1016/0167-2738(86)90195-5

[47] d'Yvoire F, Pintard-ScrGpell Mv, Bretey E, de la Rochsre M. Phase transitions and ionic conduction in 3D skeleton phosphates $A_3M_2(PO_4)_3$: A = Li, Na, Ag, K; M = Cr, Fe. Solid State Ionics. 1983;**9**(10):851-857. DOI: 10.1016/0167-2738(83)90101-7

[48] Zhang J, Yan Q, Wang F, Yuan P, Zhang P. Magnetic and transport properties of layered La_1.2Sr_1.8Mn_2-xTxO_7 (T = Fe, Co, Cr). Journal of Physics: Condensed Matter. 2000;**12**:1981-1990. DOI: 10.1088/0953-8984/12/9/303

[49] Naceur H, Megriche A, El Maaoui M. Structural distortion and dielectric properties of $Sr_{1-x}(Na_{0.5}Bi_{0.5})xBi_2Nb_2O_9$ (x = 0.0, 0.2, 0.5,0.8 and 1.0). Journal of Alloys and Compounds. 2013;**546**: 145-150. DOI: 10.1016/j.jallcom. 2012.08.101

[50] Georges S, Goutenoire F, Lacorre P, Steil M. C. Sintering and electrical conductivity in fast oxide ion conductors La_2−xRxMo_2−yWyO_9 (R: Nd, Gd, Y) Journal of the European Ceramic Society. 2005;**25**:3619-3627. DOI: 10.1016/j. jeurceramsoc.2004.09.029

[51] Georges S, Goutenoire F, Altorfer F, Sheptyakov D, Fauth F, Suard E, Lacorre P. Thermal, structural and transport properties of the fast oxide-ion conductors $La_{2-x}R_xMo_2O_9$ (R=Nd, Gd, Y). Solid State Ionics. 2003;**161**:231-241. DOI: 10.1016/S0167-2738(03)00279-0

Section 3

The Cobalt Oxide Based Composite Nanoparticles and Their Engineering and Medical Applications

Chapter 3

The Cobalt Oxide-Based Composite Nanomaterial Synthesis and Its Biomedical and Engineering Applications

Lingala Syam Sundar, Manoj K. Singh, António M.B. Pereira and Antonio C.M. Sousa

Abstract

The magnetic nanoparticles (NPs) are offering a wide range of applications in medical and engineering fields. Among all the magnetic nanoparticles, cobalt oxide (Co_3O_4) nanoparticles and its composite-based nanoparticles are attracting more interest from researchers because of its unique mechanical, thermal, and magnetic properties. The aim of this book is to bring together a number of recent contributions regarding the cobalt oxide-based composite nanoparticles from several researchers all over the world. The latest research results, innovations, and methodologies are reported in the book in order to support the discussion and to circulate ideas and knowledge about the applications of these materials in medical and engineering applications. This chapter presents the methodology for the synthesis and characterization and applications of cobalt oxide-based composite nanoparticles. The detailed analysis related to toxicity of these nanocomposite materials is also included in this book chapter.

Keywords: cobalt oxide, nanocomposite, thermal properties, electrical properties and toxicity

1. Introduction

Magnetic nanoparticles (NPs) are attracting increased researchers' interest due to their potential wide use in many engineering and medical applications. The most commonly used magnetic materials are Fe_2O_3, Fe_3O_4, Ni, Co, CoO, and Co_3O_4. However, among all these magnetic materials, cobalt oxide (Co_3O_4) nanoparticles are being preferred due to their good magnetic properties. Pure cobalt is not stable at room temperature as it can be converted to oxides like CoO, Co_2O_3, and Co_3O_4; Co_3O_4 is the most stable phase. It is a P-type semiconductor, and it has high Young's modulus, which varies between 116 and 160 GPa. Co_3O_4 exhibits a normal spinel crystal structure with occupation of tetrahedral sites by Co^{2+} and octahedral sites by Co^{3+}. Its magnetic moment arises due to Co^{2+} ions largely because of spins, with a small contribution from spin-orbit coupling. Co_3O_4 has excellent properties such as gas-sensing, catalytic, and electrochemical properties, and it has been studied widely for applications in solid-state sensors, electrochromic devices, and heterogeneous catalysts as well as lithium batteries and also medical applications [1–5]. There are several methods to synthesize the Co_3O_4 nanoparticles, which include the Co_3O_4 nanowires [6], the surfactant-templated approach for fabricating Co_3O_4 nanoboxes [7], the mechanochemical reaction method for the synthesis of Co_3O_4 nanoparticles [8], the thermal decomposition and oxidation route for the growth of Co_3O_4 nanorods [9], and the Co_3O_4 nanowalls [10].

Some of the reports dealing with the synthesis of Co_3O_4 nanoparticles and their potential use are succinctly reviewed below. Manigandan et al. [11] used the thermal decomposition method. Mariano et al. [12] synthesized Co_3O_4 nanoparticles and prepared ethylene glycol-based nanofluids. Salavati-Niasari et al. [13] prepared Co_3O_4 nanoparticles from solid organometallic molecular precursors. Salavati-Niasari et al. [14] used another method by considering benzene dicarboxylate complexes, in particular phthalate ones, as precursors. Alrehaily et al. [15] synthesized Co_3O_4 nanoparticles by gamma irradiation. All the above researchers synthesized the Co_3O_4 nanoparticles for engineering applications. Cavallo et al. [16] studied the cytotoxicity of Co_3O_4 nanoparticles in human alveolar (A549) and bronchial (BEAS-2B) cells. Alarifi et al. [17] investigated the toxicity of Co_3O_4 nanoparticles in HepG2 cells. Based on these studies, pure Co_3O_4 nanoparticles are toxic.

Cobalt-based compounds also offer interesting advantages in various applications; typical cobalt-based compounds are grapheme oxide/cobalt oxide, nanodiamond-cobalt oxide, zeolite Y/cobalt oxide, and carbon nanotubes/cobalt oxide. Syam Sundar et al. [18] synthesized GO/Co_3O_4 hybrid nanoparticles and studied their thermal properties. Syam Sundar et al. [19] also synthesized $ND\text{-}Co_3O_4$ nanoparticles and investigated their thermal properties and toxicity. Shi et al. [20] prepared different concentrations of Co_3O_4/GO, studied their catalyst activity, and observed the highest catalytic activity when the Co_3O_4 mass loading was about 50% in the catalyst. Xiang et al. [21] synthesized rGO/Co_3O_4, which was used as the pseudocapacitor electrode in the 2 M KOH aqueous electrolyte solution.

This book chapter emphasizes on the various synthesis methods for cobalt oxide and engineering and medical applications of this material. In addition, synthesis, characterization, and engineering and medical applications of cobalt oxide-based composite materials are also reviewed.

2. Cobalt oxide (Co_3O_4) nanoparticles

2.1 Synthesis procedure

There are different methods to synthesize Co_3O_4 nanoparticles, which, among others, are chemical coprecipitation, mechanochemical reaction, thermal decomposition and oxidation route, and long-time calcining method. Manigandan et al. [11] used thermal decomposition method for the synthesis of Co_3O_4 nanoparticles by dispersing 0.01 M cobalt chloride in 500 mL distilled water and 10% of glycerol. The suspension was stirred for 20 min by a magnetic stirrer at a temperature of 50°C; after that the dissolved ammonium hydroxide solution (50 mL) was added slowly to control the agglomeration. The obtained cobalt hydroxide is calcined in air for 3 h at a temperature of 450°C yielding the Co_3O_4 nanoparticles. The TEM image of the synthesized Co_3O_4 nanoparticles is shown in **Figure 1**. Salavati-Niasari et al. [13] prepared Co_3O_4 nanoparticles from a solid organometallic molecular precursor of N,N′-bis(salicylaldehyde)-1,2-phenylenediamino cobalt(II); Co(salophen) estimated the magnetic behavior of the Co_3O_4 nanoparticles. In another study, Salavati-Niasari et al. [14] used thermal deposition method for the preparation of

Co_3O_4 nanoparticles by using benzene dicarboxylate complexes, especially phthalate ones, as precursors and characterized using Fourier transform infrared and X-ray photoelectron spectroscopy and observed temperature-dependent

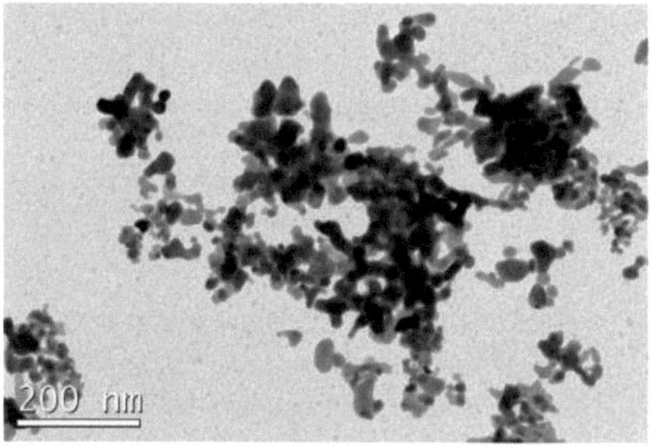

Figure 1.
TEM image of synthesized Co_3O_4 nanoparticles [12].

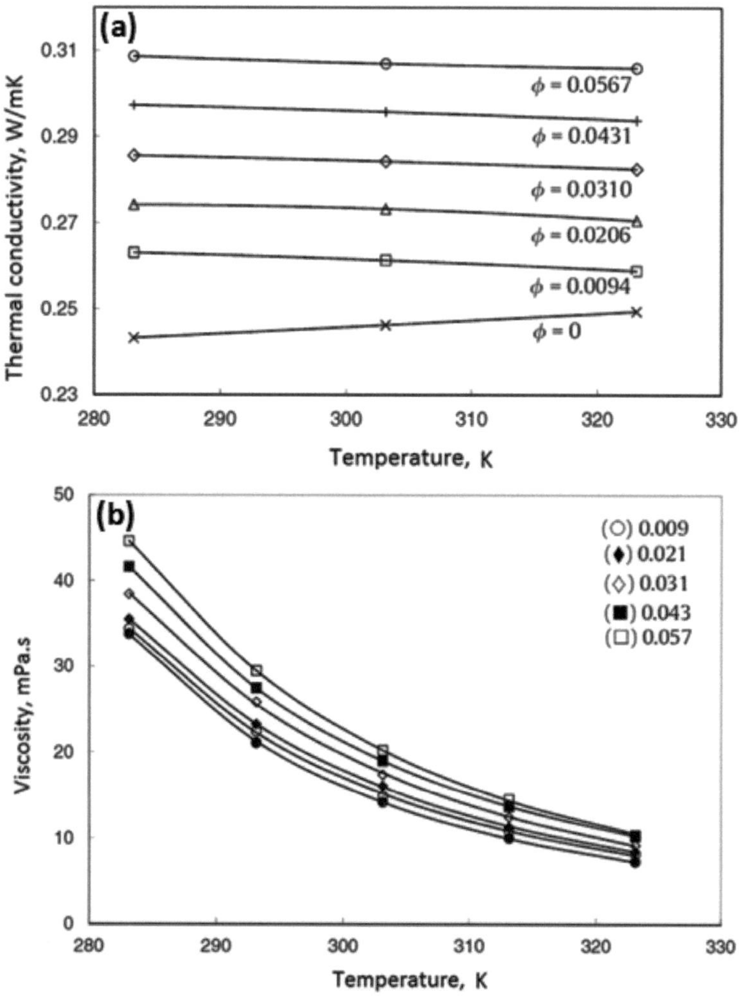

Figure 2.
Co_3O_4 nanofluids: (a) thermal conductivity and (b) viscosity [12].

magnetization curve in zero-field-cooled, where Co_3O_4 nanoparticles exhibit weak ferromagnetism. Alrehaily et al. [15] synthesized Co_3O_4 nanoparticles by gamma irradiation of 0.2–0.3 mM of $CoSO_4$ solutions. Syam Sundar et al. [18] used chemical coprecipitation method for the synthesis of Co_3O_4 nanoparticles, and they estimated their thermal properties at different particle volume concentrations and temperatures.

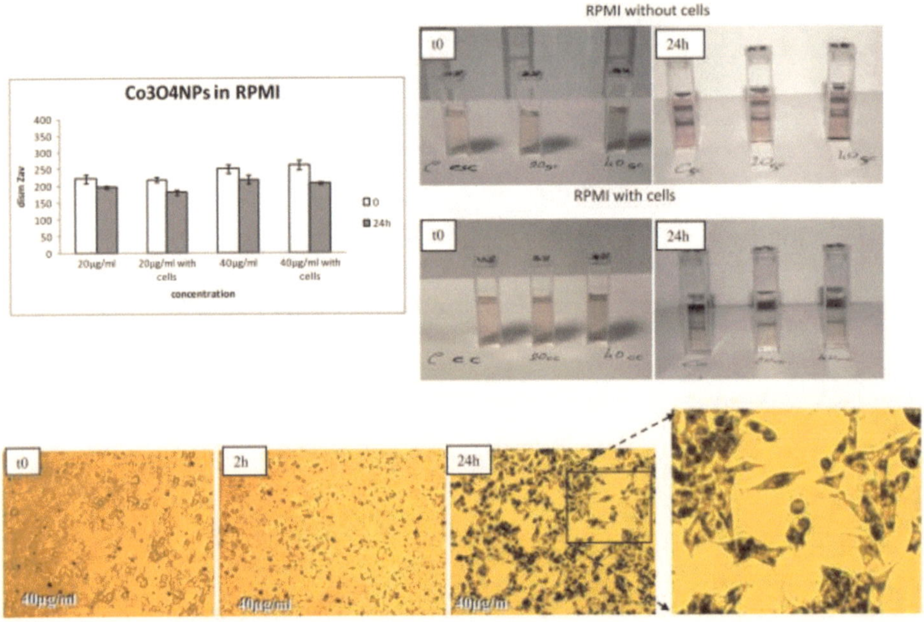

Figure 3.
Co_3O_4 nanoparticle toxicity in A549 cells: hydrodynamic size distributions of Co_3O_4 nanoparticles in RPMI 1640 medium with 10% fetal bovine serum (FBS) at t = 0 and after a 24-h exposure without and with A549 cells. In the right panels, the relative Co_3O_4 NP suspensions used for DLS measurements, showing NP sedimentation after a 24-h exposure in the medium without cells [16].

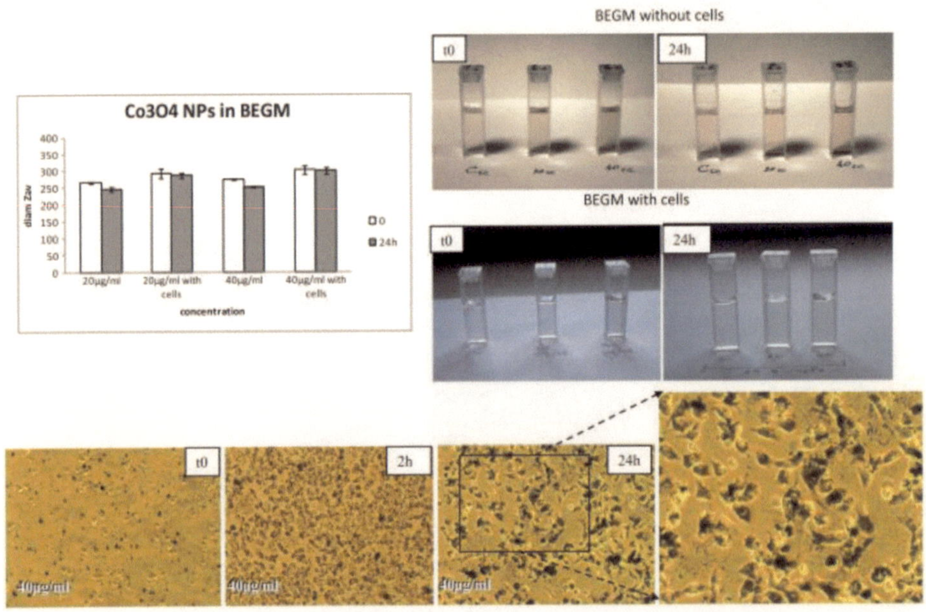

Figure 4.
Co_3O_4 nanoparticle toxicity in BEAS-2B cells: hydrodynamic size distribution of Co_3O_4 nanoparticles in BEGM medium at t = 0 and after a 24-h exposure without and with BEAS-2B cells [16].

2.2 Thermal properties

Co_3O_4 nanofluids have a potential use in several mechanical engineering applications; in particular, as replacement of low thermal conductivity fluids such as water and ethylene glycol, as a consequence, the thermal properties of nanofluids are of great interest. Mariano et al. [12] prepared Co_3O_4 nanofluids by dispersing cobalt(II, III) oxide nanopowder in ethylene glycol and determined experimentally their thermal conductivity and viscosity. The thermal conductivity and viscosity of Co_3O_4/EG nanofluids are shown in **Figure 2a** and **b**; it is noticed that the increase of particle volume concentration (ϕ) yields increased values of thermal conductivity and viscosity. They observed thermal conductivity enhancement for 5.7% volume concentration of Co_3O_4/EG nanofluids is 27% at temperature of 323.15 K (**Figure 2a**); similarly, the viscosity enhancement for 5.7% volume concentration of Co_3O_4/EG nanofluids is 40% at a temperature of 303.15 K (**Figure 2b**).

2.3 Toxicity

Knowledge about the toxicity of Co_3O_4 nanoparticles is very important considering their eventual use in medical applications. Cavallo et al. [16] studied the cytotoxicity of Co_3O_4 nanoparticles in human alveolar (A549) and bronchial

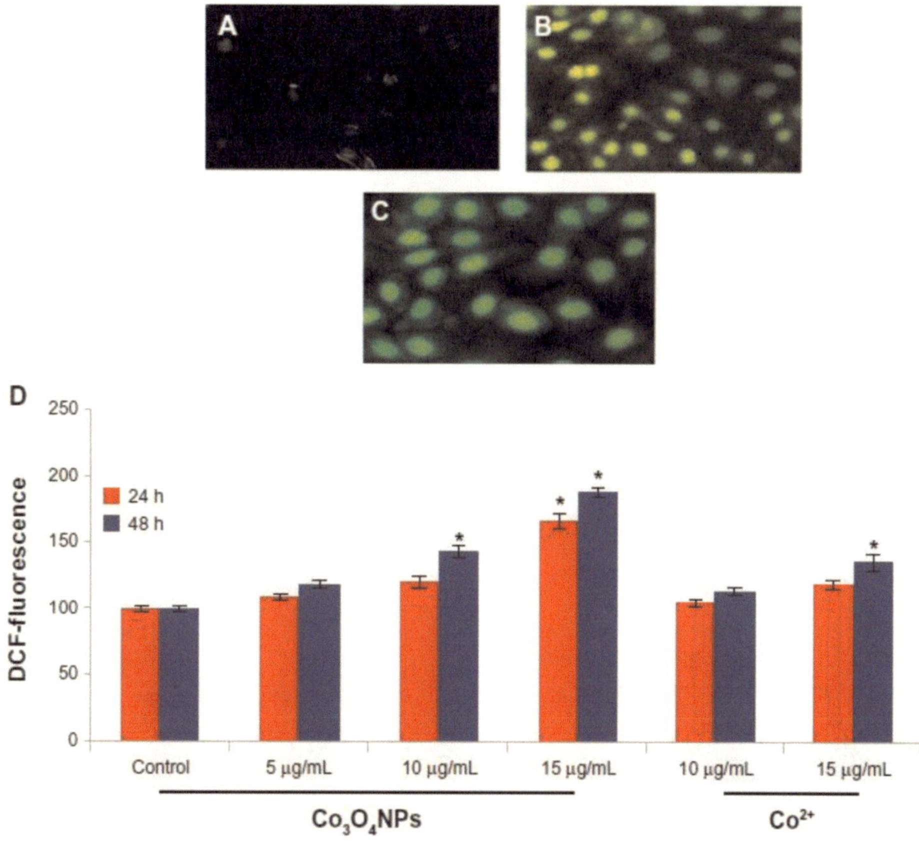

Figure 5.
Co_3O_4 nanoparticle toxicity in HepG2 cells: representative microphotographs showing Co_3O_4 NP- and Co^{2+}-induced ROS generation in HepG2 cells. Images were snapped with Nikon phase contrast with a fluorescence microscope. (A) Control, (B) 15 µg/mL of Co^{2+}, (C) 15 µg/mL of Co_3O_4NPs, and (D) percentage change in ROS generation after 24 and 48 h of exposure to various concentrations of Co_3O_4NPs and Co^{2+} in HepG2 cells [17].

(BEAS-2B) cells exposed to $1 - 40\ \mu g/mL$. In A549 cells, they found no cytotoxicity; however, BEAS-2B cells presented viability reduction at $40\ \mu g/mL$ and early membrane damage at 1, 5, and $40\ \mu g/mL$. The results related to toxicity study are presented in **Figures 3** and **4**. Alarifi et al. [17] investigated the toxicity of Co_3O_4 nanoparticles in HepG2 cells and observed cytotoxicity and genotoxicity in HepG2 cells through ROS and oxidative stress, and their results are presented in **Figure 5**.

3. Graphene oxide (GO)/cobalt oxide (Co_3O_4) nanoparticles

3.1 Synthesis procedure

The GO/Co_3O_4 composite nanoparticles were synthesized and used for various applications. Liang et al. [22] synthesized Co_3O_4/N-doped graphene hybrid nanoparticles as catalyst for oxygen reduction. They prepared GO sheet using the modified Hummers method; after the Co_3O_4/rmGO hybrid was synthesized using a $Co(OAc)_2$ aqueous solution dispersed in GO/ethanol at room temperature and stirred for 10 h at a temperature of 80°C, then NH_4OH was added to the solution. They used the GO/Co_3O_4 composite nanoparticles for catalytic activity. Syam Sundar et al. [18] synthesized the GO/Co_3O_4 hybrid nanoparticles using the chemical coprecipitation method. Their procedure involved first the preparation of GO sheets using modified Hummers method, after that, 0.2 g of GO-COOH nanosheet dispersed in 100 mL of distilled water and added the solution to 0.4 g of $CoCl_2{\cdot}6H_2O$ dispersed in 20 mL of distilled water and added 0.2932 g of $NaBH_4$, which was accompanied by the formation of a black precipitate. They prepared GO/Co_3O_4 hybrid nanofluids and observed higher values of thermal conductivity and viscosity when particle concentration and temperature increase. The synthesis method and TEM results are shown in **Figure 6**. The XRD, FTIR, and VSM results are presented in **Figure 7a–c**. The XRD patterns of GO/Co_3O_4 nanoparticles contain

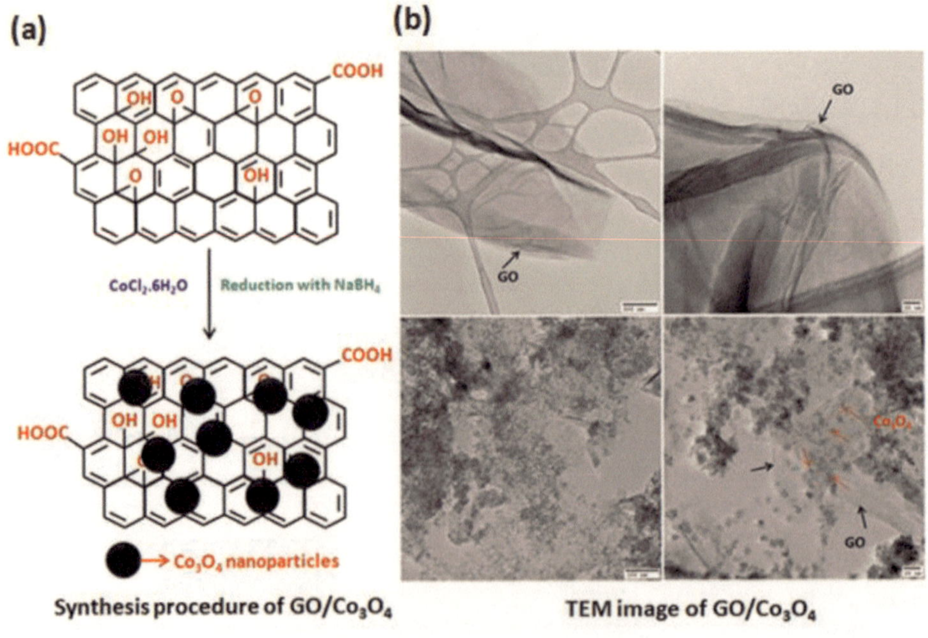

Figure 6.
GO/Co_3O_4 nanoparticles: (a) synthesis method and (b) TEM image [18].

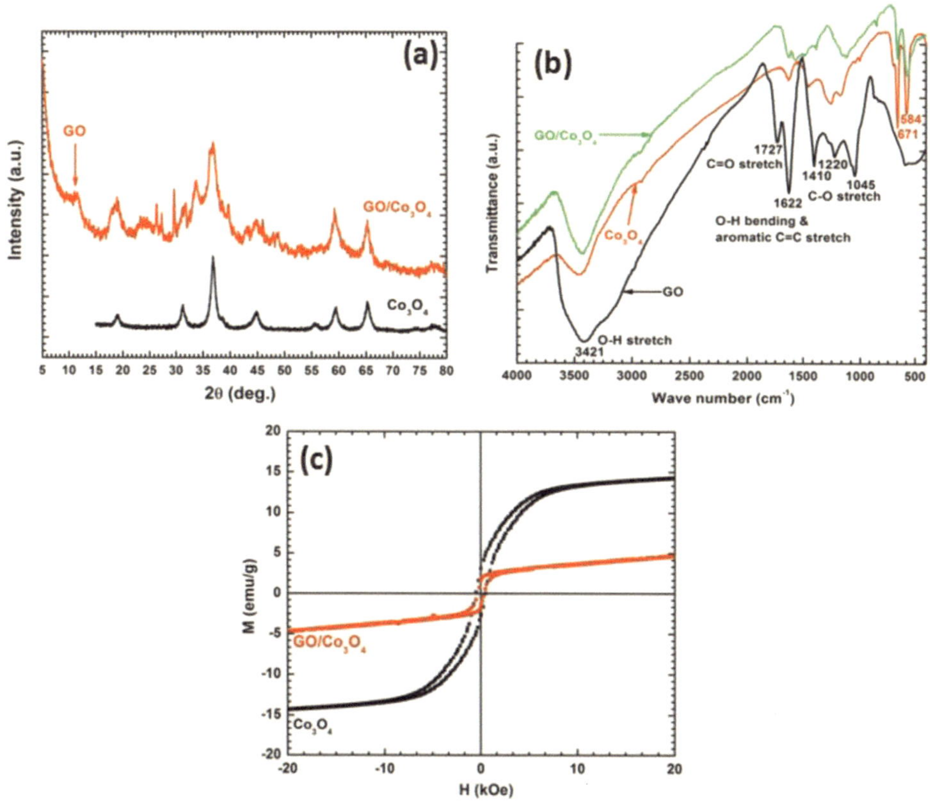

Figure 7.
GO/Co_3O_4 nanoparticles: (a) XRD patterns, (b) FTIR spectra, and (c) M-H curves [18].

both the peaks of Co_3O_4 and GO. The 2θ position of GO/Co_3O_4 nanoparticles is 11.68, 19.2, 31.2, 36.9, 44.8, 55.43, 59.4, and 65.2° which can be indexed as (002) plane for GO and (111), (220), (311), (400), (511), and (440) planes for Co_3O_4 nanoparticles (**Figure 7a**). The IR spectra of GO nanoparticle (**Figure 7c**) show the GO peak at 1622 cm^{-1} is the aromatic C=C group, the peak at 1727 cm^{-1} is the C=O vibration of carboxylic group, and the peaks at 1045 cm^{-1}, 1220 cm^{-1}, and 1410 cm^{-1} are contributed to C—O—C vibration of epoxy or alkoxy group. The IR spectra of Co_3O_4 nanoparticle show the strong band peak at 584 cm^{-1} and 671 cm^{-1} is due to the Co—O vibration of Co_3O_4 nanoparticles, which shows that Co^{2+} has been oxidized into Co_3O_4 nanoparticles. The IR spectra of GO/Co_3O_4 nanoparticles show the peaks at 1622, 1727, 1045, 1220, and 1410 cm^{-1} reflect C=C, C=O, and C—O—C groups related to GO and the peaks at 584 and 671 cm^{-1} reflect the Co—O groups related to Co_3O_4 nanoparticles. The total saturation magnetization of GO/Co_3O_4 (**Figure 7c**) is 4.67 emu/g and pure Co_3O_4 nanoparticles is 14.23 emu/g. The weight percentages of GO and Co_3O_4 nanoparticles present in the GO/Co_3O_4 nanocomposite particles were 67 and 33%, respectively. Shi et al. [20] prepared different concentrations of Co_3O_4/GO such as 20% Co_3O_4/GO, 30% Co_3O_4/GO, 50% Co_3O_4/GO, 70% Co_3O_4/GO, 90% Co_3O_4/GO, 95% Co_3O_4/GO, and pure Co_3O_4, and they studied their catalyst activity; the highest catalytic activity was observed when the Co_3O_4 loading was about 50% in the catalyst. Xiang et al. [21] synthesized 20-nm-sized Co_3O_4 nanoparticles, which were in situ grown on the chemically reduced graphene oxide (rGO) sheets to form an rGO/Co_3O_4 composite during hydrothermal processing; they were used as the pseudocapacitor electrode in the

2 M KOH aqueous electrolyte solution, and the measured specific capacitance was 472 F/g at a scan rate of 2 mV/s.

3.2 Thermal properties

The water- and ethylene glycol-based GO/Co_3O_4 nanocomposite nanofluid's thermal properties were measured by Syam Sundar et al. [18], and the results are shown in **Figures 8** and **9** at different volume concentrations and temperatures. It is noticed that the thermal conductivity of nanofluids increases linearly with the increase of particle volume concentrations and temperatures. Similarly, the viscosity of nanofluids increases with an increase of particle volume concentrations and decreases with an increase of temperature. The thermal conductivity of 0.05% nanofluid is enhanced by 2.82% and 8.58% at temperatures of 20 and 60°C, respectively, as compared to water. The thermal conductivity of 0.2% nanofluid is enhanced by 7.64 and 19.14% at temperatures of 20 and 60°C, respectively, as compared to water (**Figure 8**, left side). The viscosity of 0.05% volume concentration of nanofluid is enhanced by 1.075 times and 1.166 times; the viscosity of 0.2% volume concentration of nanofluid is enhanced by 1.49 times and 1.70 times at temperatures of 20 and 60°C compared to water (**Figure 8**, right side).

The thermal conductivity of 0.05% nanofluid is enhanced by 2.71 and 4.44% at temperatures of 20 and 60°C, respectively, as compared to EG. The thermal conductivity of 0.2% nanofluid is enhanced by 5.81 and 11.85% at temperatures of 20 and 60°C, respectively, as compared to EG (**Figure 9**, left side). The viscosity enhancement of 0.05% volume concentration of nanofluid is 1.028 times and 1.096 times; the viscosity enhancement of 0.2% volume concentration of nanofluid is 1.22 times and 1.42 times at temperatures of 20 and 60°C compared to EG (**Figure 9**, right side).

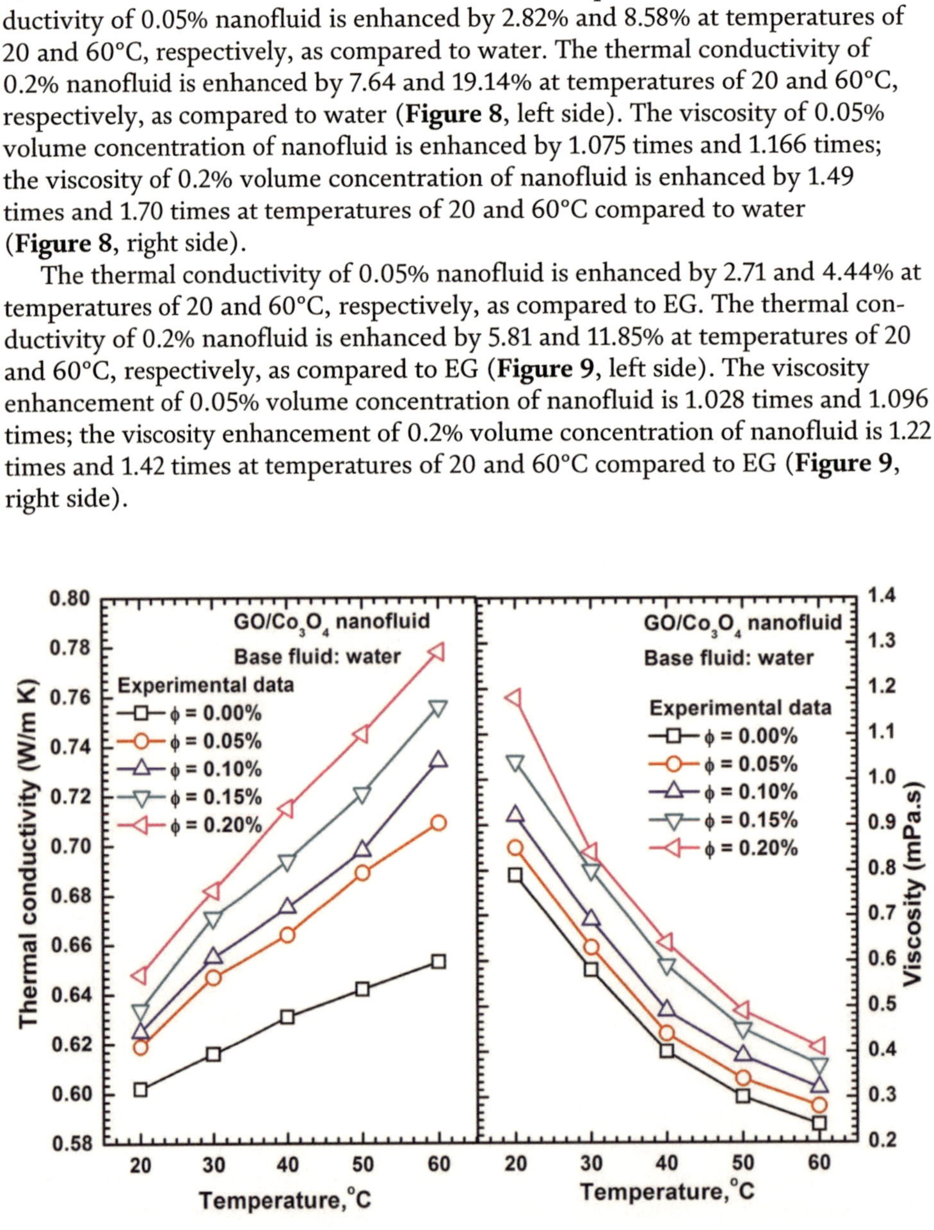

Figure 8.
Water-based GO/Co_3O_4 nanofluids: thermal conductivity (left side) and viscosity (right side) [18].

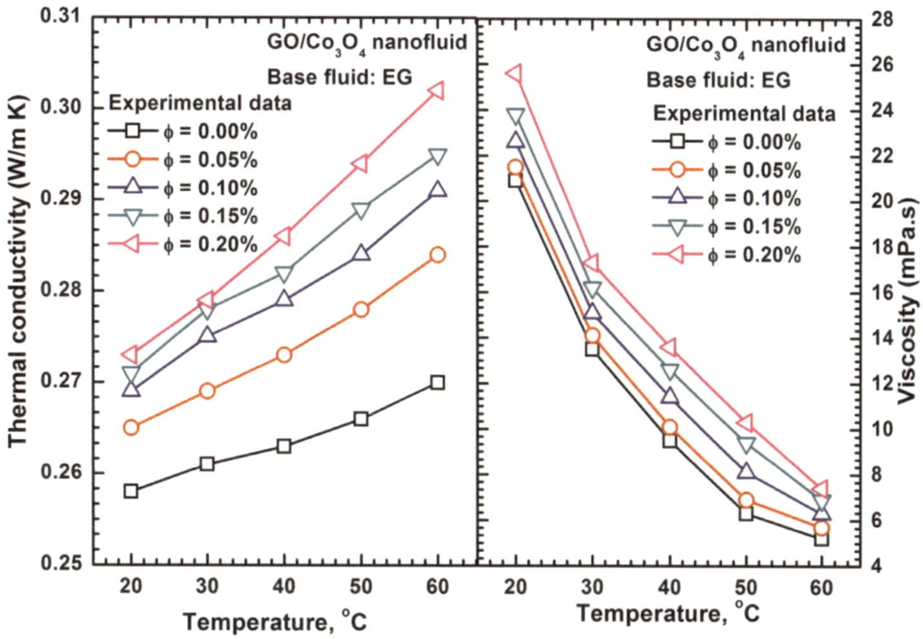

Figure 9.
Ethylene glycol-based GO/Co_3O_4 nanofluids: thermal conductivity (left side) and viscosity (right side) [18].

3.3 Electrical properties

Xiang et al. [21] measured the electrical properties of GO/Co_3O_4 using pseudocapacitor electrode in 2 M of KOH aqueous electrolyte solution. **Figure 10** depicts the galvanostatic charge discharge curve of rGO and the rGO/Co_3O_4 composite electrodes between 0 and 0.85 V at different current densities. Both the samples of rGO and rGO/Co_3O_4 electrodes exhibited good symmetric shape with the coulomb efficiency close to 1. The rGO/Co_3O_4 composite electrode presented longer charge discharge time than the rGO electrode, indicating larger specific capacitance (**Figure 10a** and **b**). The specific capacitance of the rGO/Co_3O_4 electrode was investigated with a progressively increasing current density (**Figure 10c**). The specific capacitance decreases from 458 to 416 F/g with an increase in the current density from 0.5 to 2.0 A/g, respectively. The long-term stability of the rGO/Co_3O_4 electrode was observed at the current density of 2.0 A/g. The specific capacitance of the rGO/Co_3O_4 electrode increased during the first 100 cycles, which was due to an activation process in the super capacitor electrode. About 95.6% of the specific capacitance of rGO/Co_3O_4 electrode was retained at the current density of 2.0 A/g after 1000 cycles (**Figure 10d**), which demonstrated its high cycling stability. Lai et al. [23] synthesized Co_3O_4 nanoparticles grown on nitrogen-modified microwave-exfoliated graphite oxide (NMEG) with weight ratio controlled from 10 to 70%; due to their electrochemical performance, they are used as Li-ion battery anode, and they exhibit improved cycle stability. Seventy percent of the Co_3O_4/NMEG composite has an initial irreversible capacity of 230 mAh/g (first cycle efficiency of 77%), and 910 mAh/g of capacity is retained after 100 cycles. Seventy percent of Co_3O_4/tRG-O delivers a reversible capacity of 750 mAh/g, and the irreversible capacity loss during the first cycle is 700 mAh/g. It is noted that the composite material synthesized with Co_3O_4 exhibits larger specific capacitance than rGO material when they are used as electrode. The Co_3O_4/NMEG composite material shows higher capacitance when they are used as Li-ion battery

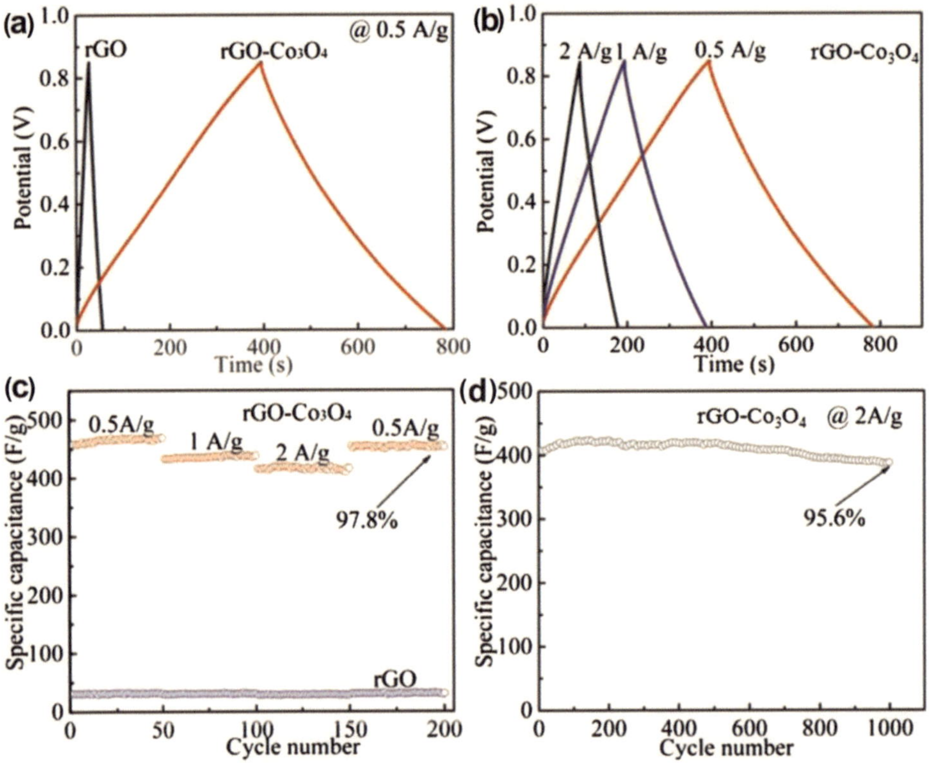

Figure 10.
GO/Co_3O_4 nanomaterial: (a) charge-discharge curves at the current density of 0.5 A/g, (b) charge-discharge curves at different current densities (0.5, 1.0, and 2.0 A/g), (c) cycling stability at varying the current density, and (d) long-term stability at a current density of 2.0 A/g [21].

anode. So, it is understood that the composite material exhibits synergistic (superior electrical) properties compared to the single-phase nanoparticles.

4. Nanodiamond (ND)-cobalt oxide (Co_3O_4) nanoparticles

4.1 Synthesis procedure

The ND-Co_3O_4 nanoparticles were synthesized by Syam Sundar et al. [19] using in situ and the chemical coprecipitation method. The synthesis route and TEM results are shown in **Figure 11**. The synthesis route contains dispersion of 0.5 g of ND particles and 0.5 g (0.003 M) of $CoCl_2{\cdot}6H_2O$ in 100 mL, adds 0.379 (0.01 M) g of $NaBH_4$ gradually, and observes the formation of light black color precipitate. The XRD, VSM, and XPS results of ND-Co_3O_4 nanocomposite are reported in **Figure 10**.

The XRD, VSM, and prepared nanofluids are shown in **Figure 12a–d**. From the XRD patterns (**Figure 12a**), the 2θ position for the ND nanoparticles for the plane (111) is 43.73°; similarly, the 2θ position for the Co_3O_4 nanoparticles for the plane (311) is 36.81°, and the ND-Co_3O_4 nanocomposite nanoparticles contains both the ND and Co_3O_4 nanoparticles planes. The weight percentage of ND and Co_3O_4 present in the ND-Co_3O_4 nanocomposite was measured from vibrating sample magnetometer (Cryogenic, UK) instrument (**Figure 12b**). The total saturation magnetization of Co_3O_4 nanoparticles is 14.3 emu/g, whereas the total saturation magnetization of ND-Co_3O_4 nanocomposite is 4.7 emu/g. It is decreased

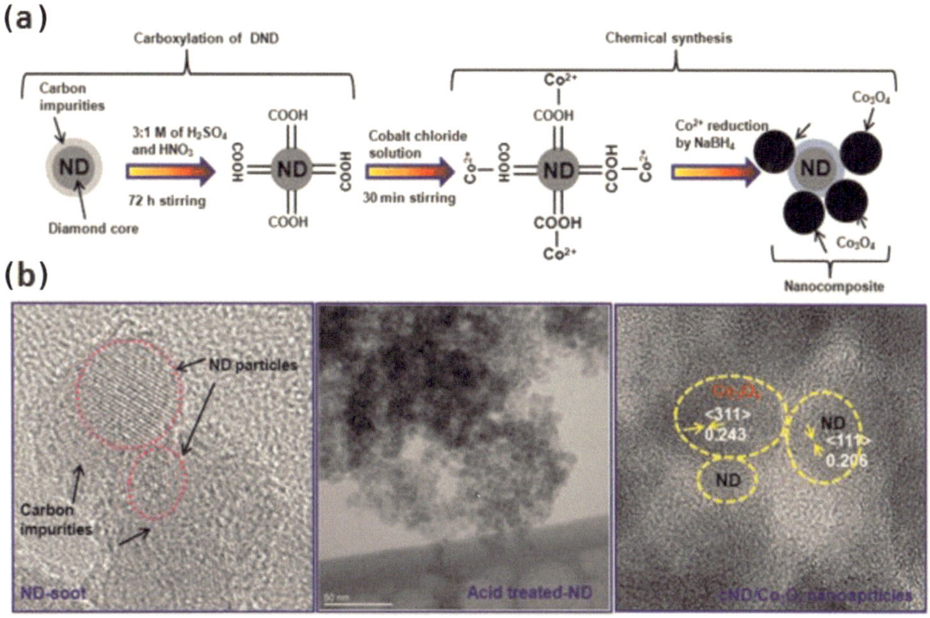

Figure 11.
ND-Co_3O_4 nanoparticles: (a) synthesized method and (b) TEM image [19].

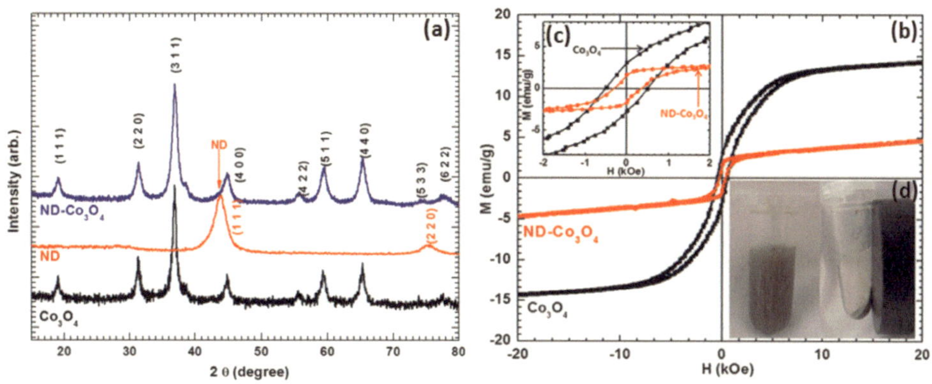

Figure 12.
ND-Co_3O_4 nanoparticles: (a) XRD patterns, (b) M–H curve, (c) coercivity, and (d) the ND-Co_3O_4 nanofluids [19].

because of the presence of nonmagnetic material of ND. The coercivity results of ND-Co_3O_4 and of pure Co_3O_4 nanoparticles are 334 and 490 Oe, respectively (**Figure 12c**). Based on the total sum rule, it is observed that there is 67% of ND and 33% of Co_3O_4 present in the ND-Co_3O_4 nanocomposite nanoparticles. The prepared ND-Co_3O_4 nanofluid's samples are shown in **Figure 12d**, and the final ND-Co_3O_4 nanocomposite particles showing magnetic behavior are observed.

The surface composition of ND-Co_3O_4 nanocomposite particles was measured using X-ray photoelectron spectroscopy (XPS), and the results are shown in **Figure 13a–c**. The Co 2p spectra has two main peaks at binding energies (BEs) of 780.7 and 796.3 eV, which can be related to Co $2p_{3/2}$ and Co $2p_{1/2}$ spin-orbit lines, respectively (**Figure 13a**). The determination of oxidation state of the each and every component is very important, and also it is very difficult. The shape of the satellites and the energy gap between the satellites are the key parameters used to discriminate between different oxidation states of Co. For instance, the presence of

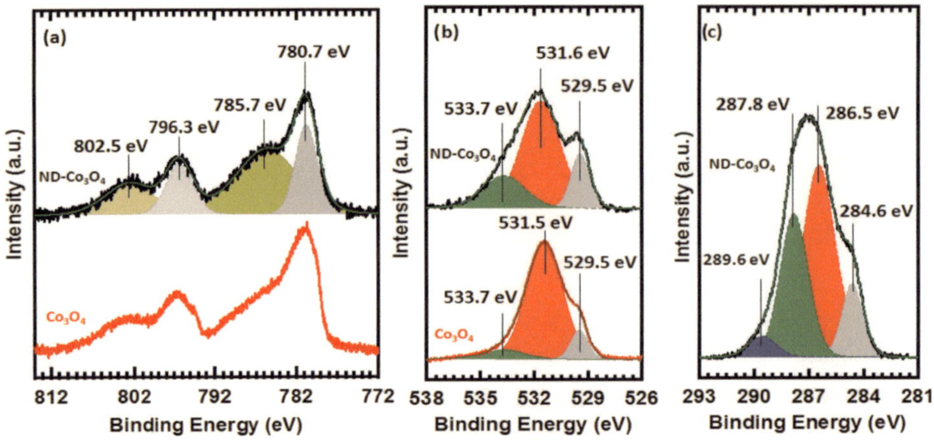

Figure 13.
ND-Co_3O_4 nanocomposite—XPS spectra: (a) Co 2P, (b) O 1s, and (c) C 1s core levels [19].

a pronounced satellite like that founded in the present sample at 785.7 eV can be ascribed to CoO. For the case of Co_3O_4 compounds, the satellite is generally detected at BEs higher than 10 eV with respect to the main peak. It is observed from the XPS analysis that the Co_3O_4 particles are covered by a thin layer of CoO.

The O 1s core level is presented in **Figure 13b** for the cobalt nanoparticles (bottom) and for the nanocomposite (upper). The first component, centered at a BE = 529.5 eV (gray), is ascribed to oxygen atoms in the cobalt particles, while the others are related to different oxygen species. In particular, the components at 531.6 eV (red) and 533.7 eV (green) are ascribed to OH^- and C—O/O=C—O, respectively. Moreover, in the case of the nanocomposite, the C 1s core level (**Figure 13c**) shows interesting features. Four components were needed for fitting this peak, appearing at BEs of 284.6 eV (gray), 286.5 eV (red), 287.8 eV (green), and 289.6 eV (blue), which can be ascribed to C—C, C—OH or C—O—C, C=O, and O=C—O, respectively. Thus, XPS indicates that the cobalt particles are integrated with the treated nanodiamonds.

4.2 Thermal properties

The water- and ethylene glycol-based ND-Co_3O_4 nanofluid's thermal conductivity and viscosity were measured by Syam Sundar et al. [19], and the data is shown in **Figure 14a** and **b** at different particle weight concentrations and temperatures. The water-based ND-Co_3O_4 nanofluid's samples are shown in **Figure 14b**, and particle size distribution is shown in **Figure 14c**. They observed thermal conductivity enhancement of 2 and 6% for 0.05 wt.% of water-based ND-Co_3O_4 nanofluid and the thermal conductivity enhancement of 8.7 and 15.7% for 0.15 wt.% of water-based ND-Co_3O_4 nanofluid at temperatures of 20 and 60°C, respectively, compared with water data (**Figure 14a**). They also observed thermal conductivity enhancement of 1.16 and 3.97% for 0.05 wt.% of EG-based ND-Co_3O_4 nanofluid and the thermal conductivity enhancement of 4.68 and 8.71% for 0.15 wt.% of EG-based ND-Co_3O_4 nanofluid at temperatures of 20 and 60°C, respectively, compared with EG data (**Figure 14d**).

4.3 Toxicity of ND-Co_3O_4 nanoparticles

The toxicity of ND-Co_3O_4 nanoparticles was studied by Syam Sundar et al. [24] on Allium cepa, and the results are shown in **Figure 15**. The untreated root tip

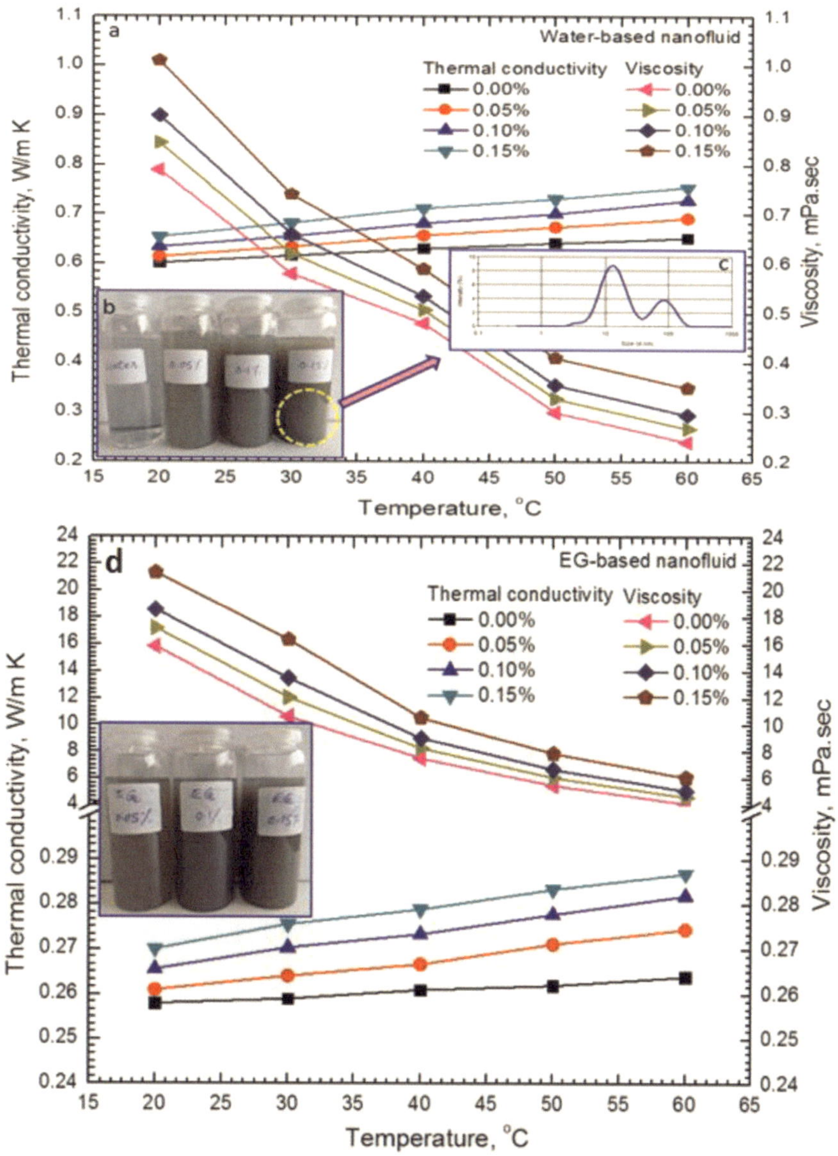

Figure 14.
Thermal conductivity of ND-Co_3O_4 nanofluids: (a) water-based nanofluids, (b) sample nanofluids, (c) particle size distribution, and (d) EG-based nanofluids [19].

cells showed a mitotic index of 71.3 ± 2.2%. However, a dose-dependent effect on mitotic index was noted for Co_3O_4 and Co_3O_4-cND. In particular, the mitotic indices were found to be 58.07 ± 1.7, 37.8 ± 1.2, and 28.6 ± 0.8% upon exposure to 5, 10, and 20 μg/mL Co, respectively. Notably, decreases in MI were insignificant at these concentrations of cND with values of 68.3 ± 2.0, 65.7 ± 1.9, and 59.0 ± 1.7, respectively. The ameliorative effect of Co-accrued impacts is demonstrated by low (5 μg/mL) and moderate (10 μg/mL) concentration values of cND-Co_3O_4. This indicates that, if accidentally released into the environment, cND-Co_3O_4 would be safe for biotic life to a maximum concentration of 10 μg/mL. The observed Co-accrued cyto-genotoxic consequences coincide with similar earlier studies, where Co oxide nanoparticles were reported to spoil the whole cellular metabolism and stages of cell division mainly by blocking water channels through

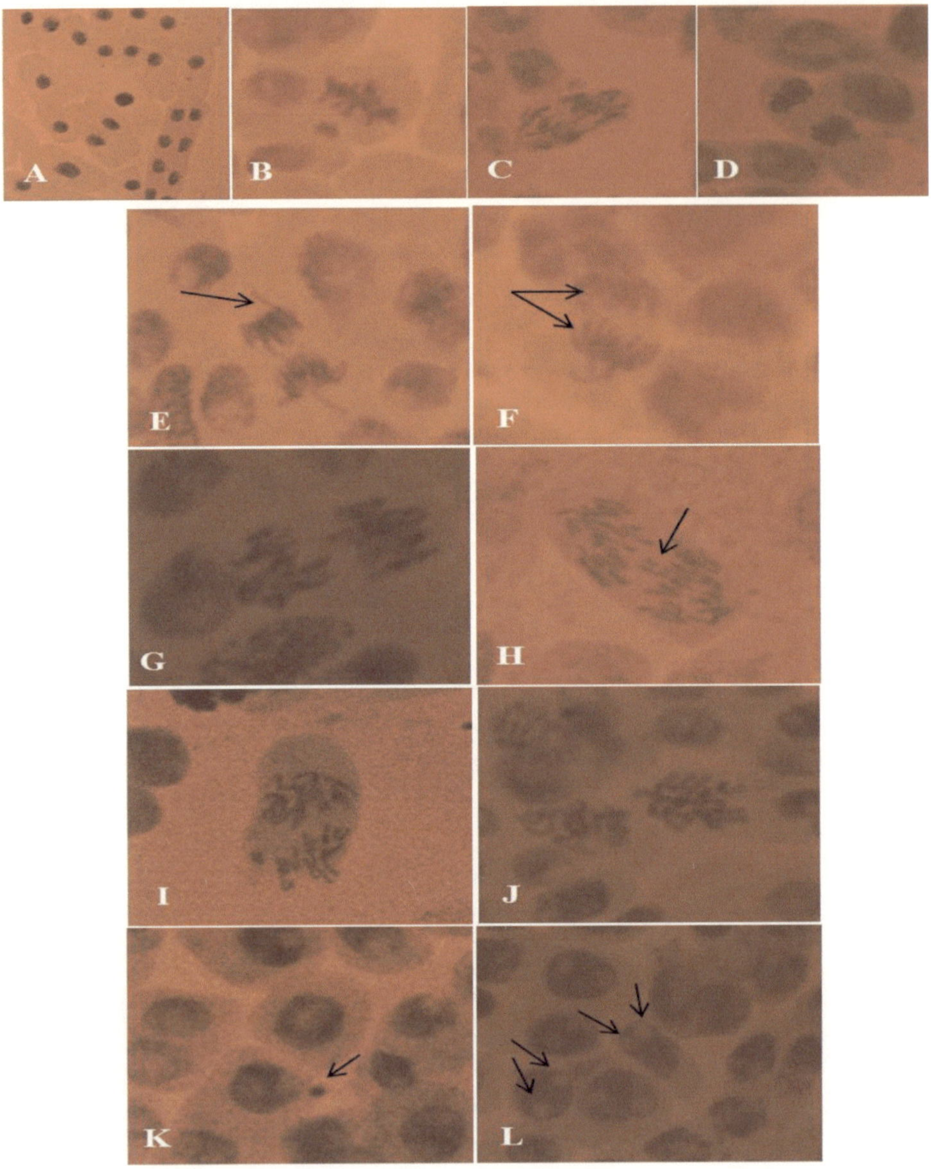

Figure 15.
Toxicity tests on Allium cepa of ND-Co_3O_4 nanoparticles (A–D), various concentrations of cobalt oxide (E–G), and cND-Co_3O_4 (H–K). A = prophase, B = metaphase, C = anaphase, D = telophase, E = chromosomal break, F = cytoplasmic bridge, G = disturbed anaphase, H = laggard, I = sticky anaphase, J = scattered anaphase, K = prophase nuclei with micronucleus in interphase, L = binucleate cells [24].

adsorption and/or by impacting genetic material by causing various types of chromosomal aberrations.

In summary, insignificant changes in MI with moderate concentration (10 μg/mL) of cND-Co_3O_4 also confirm that cND-Co_3O_4 was unable to interfere with the normal development of mitosis mainly by its incapacity to prevent cells from entering the prophase and blocking the mitotic cycle during interphase inhibiting DNA/protein synthesis. Moreover, 20 μg/mL of cND-Co_3O_4 compared to 5, 10, and 20 μg/mL of Co_3O_4 and 10 μg/mL of cND-Co_3O_4 presents insignificant and infrequent chromosome aberrations (such as stickiness, breaks, disturbed, and scattered metaphase); therefore, these results strongly support the environment-friendly nature of the cND-Co_3O_4 nanocomposite, as demonstrated by the toxicity tests conducted using Allium cepa (**Figure 15A–D**). For comparison purpose, similar tests are also performed for various concentrations of cobalt oxide nanoparticles (**Figure 15E–G**) and cND-Co_3O_4 (H–K) (**Figure 15H–K**).

5. Zeolite Y/cobalt oxide (Co_3O_4) nanoparticles

5.1 Synthesis procedure

The zeolite Y/cobalt oxide (Co_3O_4) nanoparticles were synthesized by Davar et al. [25], and the schematic diagram is depicted in **Figure 16**. The Co_3O_4 nanocomposite was synthesized by an ion exchange of cobalt ions and zeolite Y in the presence of sodium hydroxide and calcination treatment; the synthesized zeolite Y/Co_3O_4 has a paramagnetic behavior at room temperature.

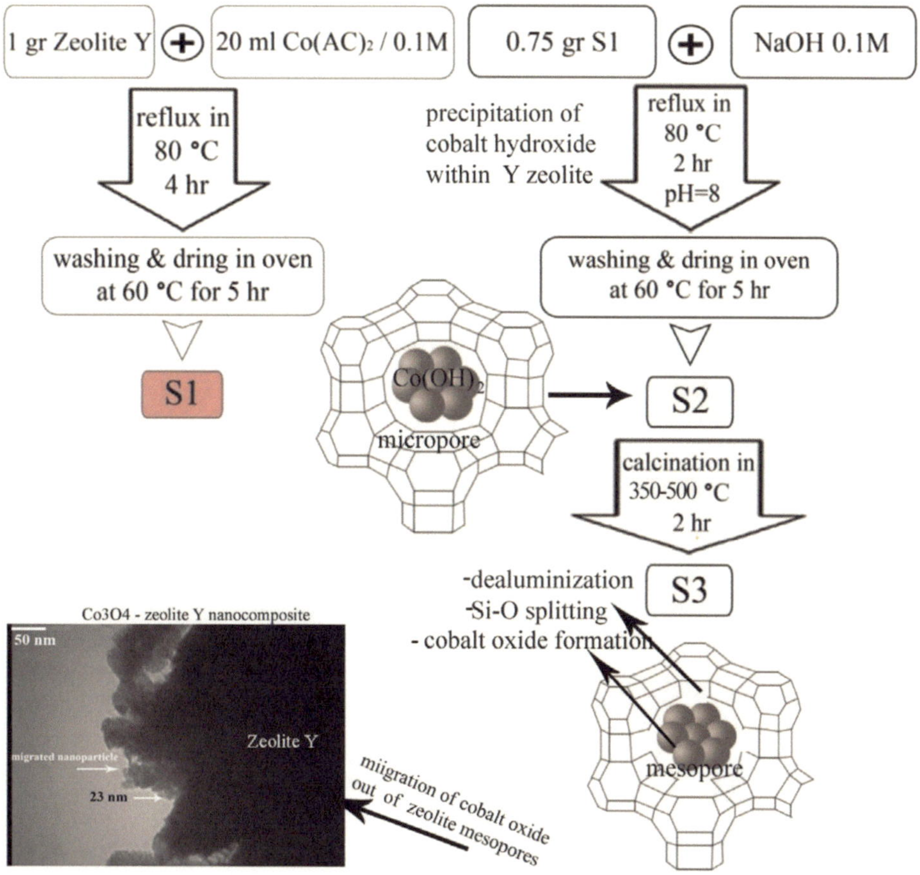

Figure 16.
Synthesis procedure of zeolite Y/Co_3O_4 nanoparticles [25].

6. Carbon nanotubes/cobalt oxide (Co_3O_4) nanoparticles

6.1 Synthesis procedure

The f-SWCNT/Co_3O_4 nanocomposite was prepared by Abdolmaleki et al. [26] using the electrostatic coprecipitation route. They noted that the specific capacitance of f-SWCNT/Co_3O_4 is high with a value of 343F/g, while that of the Co_3O_4 nanoparticles is only 77F/g (**Figure 17**). Zarnegar et al. [27] used a new sonochemical synthesis of polyhydroquinolines having as catalyst Co_3O_4-CNT nanocomposites for aldehydes, dimedone, ethyl acetoacetate, and ammonium acetate in an ethanol medium. These nanocomposites proved to be a highly effective

$$3Co(NO_3)_2 \xrightarrow[140\text{-}180\ ^\circ C]{\text{air, } n\text{-hexanol}} Co_3O_4\downarrow + 6NO_2\uparrow + O_2\uparrow$$

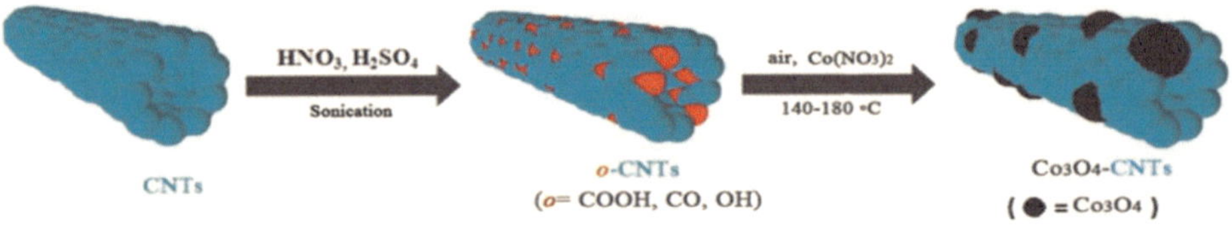

Figure 17.
Synthesis procedure of MWCNT-Co_3O_4 nanoparticles [26].

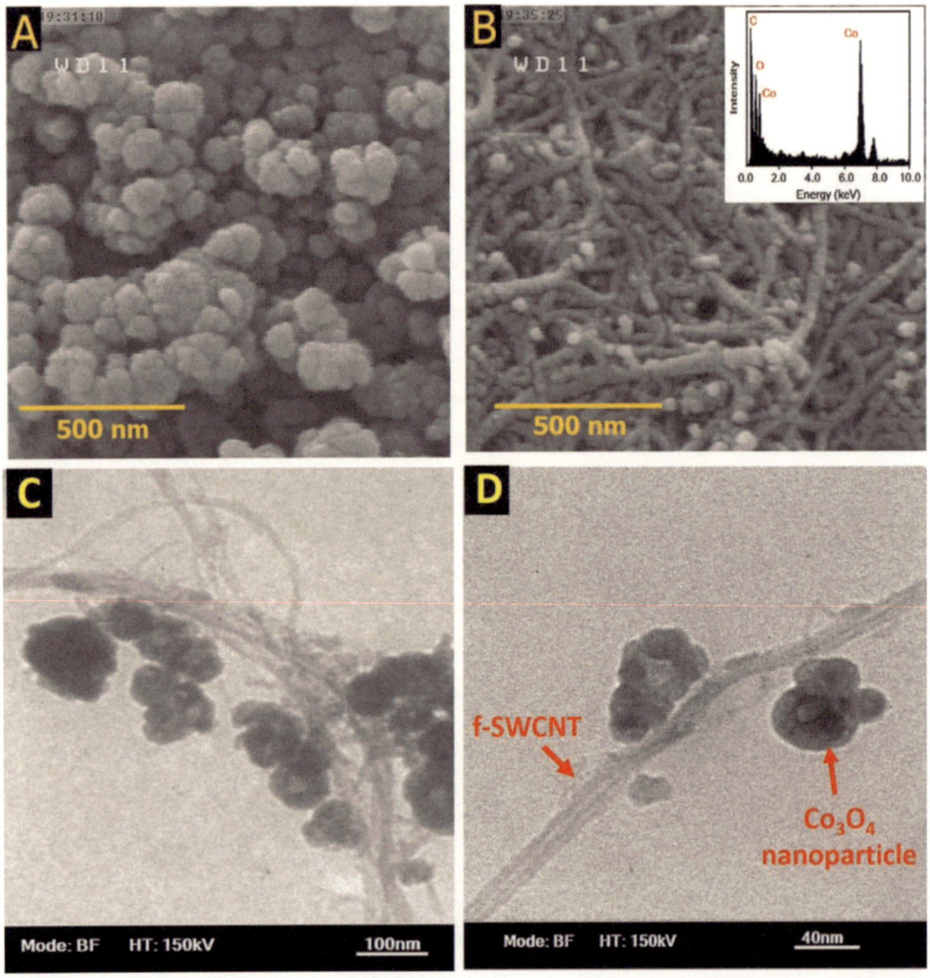

Figure 18.
TEM results of MWCNT-Co3O4 nanoparticles [27], (a) Co_3O_4, (b) MWCNT, (c) MWCNT-Co_3O_4 nanoparticles at 100 nm range and (d) MWCNT-Co_3O_4 nanoparticles at 100 nm range.

catalytic system, and they provide a green strategy to generate a variety of polyhydroquinolines under sonic conditions. The TEM image of Co_3O_4 nanoparticles (**Figure 18A**), carbon nanotubes (**Figure 18B**), Co_3O_4-CNT nanocomposites at microscope 100 nm range (**Figure 18C**), and Co_3O_4-CNT nanocomposites at microscope 40 nm range (**Figure 18D**) is shown in **Figure 18**.

7. Conclusions

The cobalt oxide-based composite nanoparticles and its engineering and medical applications were discussed in this book chapter. The pure Co_3O_4 nanoparticles reveal the presence of toxicity, whereas cobalt compounds are nontoxic. In this book chapter, the synthesis procedure for cobalt and cobalt-based compounds such as GO/Co_3O_4, ND-Co_3O_4, zeolite Y/Co_3O_4, and MWCNT-Co_3O_4. All the synthesized cobalt-based compound materials present magnetic properties, which can be benefit for the thermal applications, electrical applications, and medical applications. Cobalt oxide-based compounds have improved synergistic properties as compared to pure cobalt oxide nanoparticles.

Acknowledgements

The author (LSS) acknowledges the Foundation for Science and Technology (FCT, Portugal) for the financial support received through the grant SFRH/BPD/100003/2014. The author (ACMS) acknowledges the 2017 Visiting Scientist Fellowship awarded to him under the Chinese Academy of Sciences President's International Fellowship Initiative. TEMA/DEM researchers also acknowledge the FCT grant UID/EMS/00481/2019-FCT, the infrastructures support CENTRO-01-0145-FEDER-022083-Centro Portugal Regional Operational Programme (Centro2020), and Project 33912-AAC no. 03/SI/2017, under the Portugal 2020 Partnership Agreement, through the European Regional Development Fund.

Author details

Lingala Syam Sundar[1]*, Manoj K. Singh[1,2]*, António M.B. Pereira[1] and Antonio C.M. Sousa[1]

1 Department of Mechanical Engineering, Centre for Mechanical Technology and Automation (TEMA-UA), University of Aveiro, Portugal

2 Centre for Nano and Material Sciences, JAIN (Deemed-to-be University), Bengaluru, Karnataka, India

*Address all correspondence to: sslingala@gmail.com and mksingh@ua.pt

References

[1] Alinovi R, Goldoni M, Pinelli S, Campanini M, Aliatis I, Bersani D, et al. Oxidative and pro inflammatory effects of cobalt and titanium oxide nanoparticles on aortic and venous endothelial cells. Toxicology In Vitro. 2014;**29**:426-437

[2] Cattaneo AG, Gornati R, Sabbioni E, Chiriva-Internati M, Cobos E, Jenkins MR, et al. Nanotechnology and human health: Risks and benefits. Journal of Applied Toxicology. 2010;**30**: 730-744

[3] Chattopadhyay S, Dash SK, Tripathy S, Das B, Mandal D, Pramanik P, et al. Toxicity of cobalt oxide nanoparticles to normal cells; an in vitro and in vivo study. Chemico-Biological Interactions. 2014;**226**:58-71

[4] Cho WS, Duffin R, Bradley M, Megson IL, Macnee W, Howie SE, et al. NiO and Co_3O_4 nanoparticles induce lung DTH-like responses and alveolar lipoproteinosis. The European Respiratory Journal. 2012;**39**:546-557

[5] Cho WS, Duffin R, Bradley M, Megson IL, MacNee W, Lee JK, et al. Predictive value of in vitro assays depends on the mechanism of toxicity of metal oxide nanoparticles. Particle and Fibre Toxicology. 2013;**10**:55

[6] Zhu Y, Stubbs LP, Ho F, Liu R, Ship CP, Maguire JA, et al. Magnetic nanocomposites: A new perspective in catalysis. ChemCatChem. 2010;**2**: 365-374

[7] Lim CW, Lee IS. Magnetically recyclable nanocatalyst systems for the organic reactions. Nano Today. 2010;**5**: 412-434

[8] Chen GJ, Wang LF. Design of magnetic nanoparticles-assisted drug delivery system. Current Pharmaceutical Design. 2011;**17**: 2331-2351

[9] Wan TJ, Shen SM, Siao SH, Huang CF, Cheng CY. Using magnetic seeds to improve the aggregation and precipitation of nanoparticles from backside grinding wastewater. Water Research. 2011;**45**:6301-6307

[10] Yang HT, Su YK, Shen CM, Yang TZ, Goa HJ. Synthesis and magnetic properties of ε-cobalt nanoparticles, Surface and Interface Analysis. 2004; **36**:155-160

[11] Manigandan R, Giribabu K, Suresh R, Vijayalakshmi L, Stephen A, Narayanana V. Cobalt oxide nanoparticles: Characterization and its electrocatalytic activity towards nitrobenzene. Chemical Science Transactions. 2013;**2**(S1):S47-S50

[12] Mariano A, Pastoriza-Gallego MJ, Lugo L, Mussari L, Piñeiro MM. Co_3O_4 ethylene glycol-based nanofluids: Thermal conductivity, viscosity and high pressure density. International Journal of Heat and Mass Transfer. 2015; **85**:54-60

[13] Salavati-Niasari M, Khansari A, Davar F. Synthesis and characterization of cobalt oxide nanoparticles by thermal treatment process. Inorganica Chimica Acta. 2009;**362**:4937-4942

[14] Salavati-Niasari M, Mir N, Davar F. Synthesis and characterization of Co_3O_4 nanorods by thermal decomposition of cobalt oxalate. Journal of Physics and Chemistry of Solids. 2009;**70**:847-852

[15] Alrehaily LM, Joseph JM, Biesinger MC, Guzonas DA, Wren JC. Gamma-radiolysis-assisted cobalt oxide nanoparticle formation. Physical Chemistry Chemical Physics. 2013;**15**: 1014

[16] Cavallo D, Ciervo A, Fresegna AM, Maiello R, Tassone P, Buresti G, et al. Investigation on cobalt-oxide nanoparticles cyto-genotoxicity and inflammatory response in two types of respiratory cells. Journal of Applied Toxicology. 2015;**35**:1102-1113

[17] Alarifi S, Ali D, Suliman AO, Ahamed M, Siddiqui MA, Al-Khedhairy AA. Oxidative stress contributes to cobalt oxide nanoparticles-induced cytotoxicity and DNA damage in human hepatocarcinoma cells. International Journal of Nanomedicine. 2013;**8**: 189-199

[18] Syam Sundar L, Singh MK, Ferro MC, Sousa ACM. Experimental investigation of the thermal transport properties of graphene oxide/Co_3O_4 hybrid nanofluids. International Communications in Heat and Mass Transfer. 2017;**84**:1-10

[19] Syam Sundar L, Irurueta GO, Venkata Ramana E, Manoj K, Singh ACMS. Thermal conductivity and viscosity of hybrid nanofluids prepared with magnetic nanodiamond-cobalt oxide (ND-Co_3O_4) nanocomposite. Case Studies in Thermal Engineering. 2016;**7**: 66-77

[20] Shi P, Dai X, Zheng H, Li D, Yao W, Hu C. Synergistic catalysis of Co_3O_4 and graphene oxide on Co_3O_4/GO catalysts for degradation of Orange II in water by advanced oxidation technology based on sulfate radicals. Chemical Engineering Journal. 2014;**240**:264-270

[21] Xiang C, Li M, Zhi M, Manivannan A, Wu N. A reduced graphene oxide/Co_3O_4 composite for supercapacitor electrode. Journal of Power Sources. 2013;**226**:65-70

[22] Liang Y, Li Y, Wang H, Zhou J, Wang J, Regier T, et al. Co_3O_4 nanocrystals on graphene as a synergistic catalyst for oxygen reduction reaction. Nature Materials. 2011;**10**: 780-786

[23] Lai L, Zhu J, Li Z, Yu DYW, Jiang S, Cai X, et al. Co_3O_4/nitrogen modified graphene electrode as Li-ion battery anode with high reversible capacity and improved initial cycle performance. Nano Energy. 2014;**3**:134-143

[24] Syam Sundar L, Anjum NA, Ferro MC, Pereira E, Singh MK, Sousa ACM. Biocompatibility and biotoxicity of in-situ synthesized carboxylated nanodiamond-cobalt oxide nanocomposite. Journal of Materials Science and Technology. 2017;**33**: 879-888

[25] Davar F, Fereshteh Z, Razavi HS, Razavi RS, Loghman-Estarki MR. Synthesis and characterization of cobalt oxide nanocomposite based on the Co_3O_4-zeolite Y. Superlattices and Microstructures. 2014;**66**:85-95

[26] Abdolmaleki A, Kazerooni H, Gholivand MB, Heydari H, Pendashteh A. Facile electrostatic coprecipitation of f-SWCNT/Co_3O_4 nanocomposite as supercapacitor material. Ionics. 2015;**21**:515-523

[27] Zarnegar Z, Safari J, Kafroudi ZM. Co_3O_4-CNT nanocomposites: A powerful, reusable, and stable catalyst for sonochemical synthesis of polyhydroquinolines. New Journal of Chemistry. 2015;**39**:1445-1451

Section 4

Cobalt Based Catalysts for CO Preferential Oxidation

Chapter 4

Cobalt-Based Catalysts for CO Preferential Oxidation

Leticia E. Gómez and Alicia V. Boix

Abstract

In this work, catalysts based on cobalt supported on ZrO_2 and CeO_2 and CoCeMnOx were studied for the CO preferential oxidation (COPrOx) in hydrogen-rich stream able to feed fuel cells. Among them, the CoCeMnOx formulation showed the highest CO conversion at low temperatures, while the cobalt oxide supported on ceria presented the best selectivity toward CO_2. The Co_3O_4 spinel was the active phase for the CO preferential oxidation detected in all catalysts. However, the CoOx-CeO_2 and CoCeMnOx catalysts resulted more active than cobalt oxide supported on zirconia. The presence of ceria close to cobalt species promotes the redox properties and enhances the catalytic activity. In the CoCeMnOx catalyst prepared by coprecipitation, the incorporation of Mn represented an additional positive effect. The presence of Mn promoted the reoxidation of Co^{2+} to Co^{3+} and, consequently, the activity increased at low temperature. By X-ray diffraction (XRD) of CoOx-ZrO_2 and the CoOx-CeO_2 catalysts, the Co_3O_4 spinel and ZrO_2 or CeO_2 were identified in agreement with laser-Raman spectra. At the same time, the CoCeMnOx catalyst, prepared by coprecipitation of precursor salts, showed an incipient development of a new phase $(Mn,Co)_3O_4$ mixed spinel, due to the intimate contact between elements.

Keywords: COPrOx, $(Mn,Co)_3O_4$ mixed spinel, redox couple, CoCeMnOx, CeO_2 support, XPS, laser-Raman spectroscopy

1. Introduction

The global demand for energy has been inexorably growing in the last decades. The increasing use of fossil fuels in order to generate energy causes serious problems to the environment due to the gaseous emissions. This fact has produced a global movement toward trying to remove the contaminants from combustion effluents. Besides, the exploration for alternatives to fossil fuels, biofuels or hydrogen as an energy vector, has gained an immediate and future significance because this could contribute to the depletion of greenhouse gases [1–3].

Fuel cells are devices that are being actively developed, because they are power generation systems that can produce energy with significantly less impact on the environment. Among several types of fuel cells, proton exchange membrane fuel cells (H_2-PEMC) are considered to be the most technically advanced for such application [4]. Hydrogen produced by means of the steam reforming or auto-thermal process of hydrocarbons or alcohols should contain less than 10 ppm of carbon monoxide before entering the cell since CO poisons the Pt anode of the fuel cells (**Figure 1**) [5, 6]. After the reforming step, the hydrogen production process

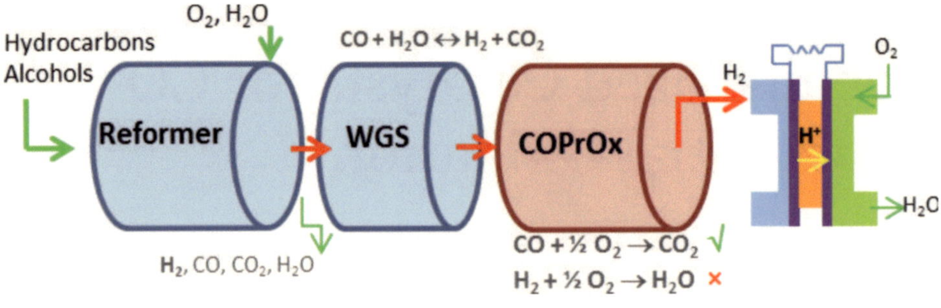

Figure 1.
Schematic flow diagram of a typical fuel cell processor with COPrOx.

continues with the water gas shift reaction (WGSR), where the stream is enriched in hydrogen and the CO content is diminished to 1%. As the CO concentration in the hydrogen stream is so high to enter to the cell, it is necessary to reduce it to the desired levels. Among various methods such as catalytic methanation, Pd-based membrane, and catalytic CO preferential oxidation (COPrOx), the latter is considered as the most adequate due to its simplicity and effectiveness. Thus, given the importance of the hydrogen purification process, the last few years have witnessed the surge of a renewed interest in the CO oxidation reaction, and several contributions dealing with this issue have recently been published [5–7].

In our research group, catalysts for the removal of contaminants from vehicles, industrial facilities, and power-generating sources have been studied [8, 9]. In addition, catalysts and catalytic reactors have also been investigated for the production and purification of H_2 to be used in fuel cells [10–13].

On the other hand, due to their redox properties, cobalt-containing catalysts have been the object of numerous publications in the environmental catalysis field and in the purification of H_2 stream, among other applications. For example, catalysts based on cobalt oxides have been studied for soot combustion [14], NOx selective reduction in oxygen excess [15], and CO preferential oxidation reaction [16].

The challenges involved in the design of successful COPrOx catalysts are the following: (i) high CO oxidation activity at low temperatures, (ii) high selectivity for CO oxidation against the undesired H_2 oxidation, and (iii) good resistance to deactivation caused by the H_2O and CO_2 in the feed [5, 6, 17].

In order to reduce the CO concentration without consuming H_2, several catalysts have been studied with promising results. Catalysts of Pt supported on bimetallic systems achieved excellent results [17–19], but the high cost of noble metals led the investigations toward the use of other active phases. For instance, CuO/CeOx catalysts have shown a catalytic performance in the COPrOx comparable to those based on noble catalysts [20–22].

In this vein, it has been shown that the combined effect of cobalt oxide and ceria strongly influences the morphological and redox properties of the composite oxides, by dispersing the Co_3O_4 phase and promoting the efficiency of the Co^{3+}/Co^{2+} redox couple [23, 24]. Numerous publications have shown that the addition of other metals to a cerium-based mixed oxide (Zr or Mn) increases the oxygen storage capability of the ceria. This effect is originated by the increase in the surface oxygen mobility due to chemical interactions between cobalt, cerium, and other metals [25–27].

On the other hand, the use of structured catalysts as those supported onto honeycomb monoliths has long been considered in the chemical industry, and it has increased with the significance of environmental catalysis in pollution abatement applications due to catalysts with a high attrition resistance and a low pressure drop that are required. In addition the thin catalytic coating allows high efficiency and selectivity. Cordierite offers high mechanical strength, high resistance to elevated temperatures, and temperature shocks due to its low thermal expansion coefficient. Moreover, it presents great adhesion stability, which is a very important property for the monolithic catalysts [12, 13, 15, 16].

In our previous work with structured catalysts, we showed that cobalt supported on zirconia and ceria and CoCeMnOx mixed oxides resulted in efficient catalysts for the COPrOx. However, it is of great interest both to identify the cobalt species present in an active catalyst and to analyze the interactions with the support (CeO_2 or ZrO_2) and their influence on the catalytic behavior. It is also important to understand the role of cobalt in the CoCeMnOx mixed oxides. In the present work, a very detailed analysis of the species present in these complex systems was carried out. The catalysts were analyzed by X-ray diffraction (XRD) and temperature-programmed reduction (TPR). The X-ray photoelectron spectroscopy (XPS) and laser-Raman spectroscopy (LRS) were used to characterize the oxidation state, concentration, and chemical nature of species present on catalysts.

2. Materials and methods

2.1 Preparation of cobalt-based catalysts

Three main groups of Co-based catalysts with 10 wt.% of cobalt were prepared—CoOx/ZrO_2, CoOx-CeO_2, and CoCeMnOx—in order to study the cobalt species and their relation to the centers active in the CO preferential oxidation.

The CoOx-ZrO_2 catalyst with 10 wt.% cobalt on ZrO_2 was prepared by wet impregnation and labeled CoZ. A commercial zirconia support (6.8 m^2/g, pore volume ca. 0.013 cm^3/g) was impregnated with an aqueous solution of $Co(NO_3)_2$. The mixture was evaporated until achieving a paste, which was placed in the oven overnight at 120°C.

CoOx-CeO_2 catalysts were prepared by two different methods. The CoCe-I sample was produced by the wet impregnation method with 10 wt.% of cobalt, using an aqueous solution of $Co(NO_3)_2$ and CeO_2 powder as support which was obtained by precipitation of the $Ce(NO_3)_3$ solution with $NH_4(OH)$. The mixture was evaporated under continuous agitation until achieving a paste, which was placed in the oven overnight at 110°C. By means of the coprecipitation method, CoCe-P catalyst was obtained. The aqueous solution of $Co(NO_3)_2$ and $Ce(NO_3)_3$ was precipitated with $NH_4(OH)$ added drop by drop under vigorous stirring. The resulting precipitate was filtered and washed several times with distilled water and dried overnight at 110°C.

The CoCeMnOx oxide mixture catalyst with 10 wt.% of Co and Mn/Co molar ratio of 1/4 was prepared by the coprecipitation method, adding $NH_4(OH)$ to an aqueous solution of $Co(NO_3)_2$, $Ce(NO_3)_3$, and $Mn(NO_3)_2$. The mixture was kept 2 h under continuous stirring at room temperature. The precipitate obtained was washed several times with deionized water and then dried overnight at 110°C.

After drying, all obtained solids were calcined for 5 h at 500°C under flowing air (25 mL/min), and the temperature was ramped from room temperature up to 500°C at 10°C/min.

2.2 Characterization

2.2.1 Chemical composition quantification

Elemental analyzes were performed by inductively coupled plasma atomic emission spectroscopy (ICP-AES) on an ICP Optima 2100 DV PerkinElmer instrument.

2.2.2 Surface area measurement

The specific surface area was calculated by BET method from N_2 adsorption isotherms at −196°C with a Micromeritics TriStar 3000 instrument. The powder samples (50 mg) were previously outgassed in vacuum at 300°C for 3 h in order to remove any adsorbed substance. The volume of adsorbed N_2 was measured as the pressure was increased.

The surface area S of the catalysts was calculated as

$$S = \frac{v_m N A}{VM} \tag{1}$$

where N is the Avogadro number, VM is the molar volume of a gas, A is the area of a molecule of nitrogen, and v_m is the volume of an adsorbed monolayer of gas, which is obtained from the BET equation

$$\frac{P}{v(P_0 - P)} = \frac{1}{v_m c} + \frac{(c-1)}{v_m}\frac{P}{P_0} \tag{2}$$

where P and P_0 are the equilibrium and the saturation pressure of N_2 at the temperature of adsorption, v is the adsorbed gas volume, v_m is the volume of an adsorbed monolayer, and c is the BET constant.

2.2.3 X-ray diffraction (XRD)

The patterns of all samples were measured on a Shimadzu XD-D1 instrument with monochromator using Cu-Kα radiation at a scanning rate of 1°/min in the range of 2θ = 20°–80°.

2.2.4 Temperature-programmed reduction (H_2-TPR)

The catalysts (100 mg) were pretreated in Ar flow heating up to 300°C and held 30 min in order to clean the catalytic surface. The reducing gas was a 5% H_2/Ar mixture; the temperature was ramped up at 10°C/min to 900°C. The experiments were carried out using an Okhura TP-2002S instrument equipped with a thermal conductivity detector (TCD). The H_2 uptake was registered as the temperature increased.

2.2.5 Laser-Raman spectroscopy (LRS)

The laser-Raman spectra of catalysts and CeO_2 and ZrO_2 supports disposed as pressed powder were measured with a LabRAM spectrometer (HORIBA Jobin Yvon) with an Olympus confocal microscope, equipped with a charge-coupled device (CCD) detector. The excitation wavelength was in all cases 532 nm. The laser power was set at 30 mW. In addition, a Linkam high-temperature cell was used for the in situ experiments, introducing into the cell 50 mg of powder catalysts. The reductive/oxidative treatments were carried out at different temperatures. The concentration and the caudal of gas reactants which entered to the cell were controlled by mass flow controllers. A thermocouple placed on the middle of the catalytic cell allowed the temperature control.

2.2.6 X-ray photoelectron spectroscopy

XPS analyses were performed in a multi-technique system (SPECS) equipped with AlKα X-ray source and operated at 100 W with pass energy of 30 eV. In addition, in order to avoid the spectral interference (Mn LMM with Ce 3d), the spectra of Ce 3d, O 1s, and C 1s were measured with MgKα X-ray source. The XPS analyses were performed on the calcined powders which were previously pressed. The data treatment was performed with the Casa XPS program (Casa Software Ltd., UK).

2.3 Assessment of the catalytic activity

CO preferential oxidation experiments were performed in a fixed-bed flow reactor at atmospheric pressure. The reaction mixture consisted of 1% CO, 1% O_2, and 40% H_2 and He balance. The weight/total flow ratio was adjusted to 2.1 mg•min/mL. The CO conversion (X_{CO}) and the oxygen selectivity toward CO_2 (S_{O2}) were defined as

$$X_{CO}(\%) = \frac{(C_{CO}^{0} - C_{CO})}{C_{CO}^{0}} \times 100 \tag{3}$$

$$S_{O_2}(\%) = 0.5\frac{(C_{CO}^{0} - C_{CO})}{(C_{O_2}^{0} - C_{O_2})} \times 100 \tag{4}$$

where C_{CO} and C_{O2} were reactor exit concentrations of CO and O_2, respectively, while $C_{CO}{}^{0}$ and $C_{O2}{}^{0}$ represented the feed concentrations. The stream compositions were measured with a chromatograph GC-2014 Shimadzu equipped with a TCD cell. All catalysts were pretreated during 1 h in O_2/He flow at 200°C before the catalytic test. In order to study the stability and a possible deactivation caused by the presence of CO_2, the CoCeMnOx catalyst was tested during 75 h at 175°C, adding 20% CO_2 in the reactant mixture.

3. Results and discussion

3.1 Prepared catalysts

The compositions of the cobalt-based catalysts, analyzed by the ICP technique, are presented in **Table 1** and are in agreement with the predefined nominal values. The cobalt content (wt.%) resulted 9.2, 10.6, and 8.6 for CoZ, CoCe-I, and CoCe-P, respectively, while for the mixture of oxides CoCeMnOx, the concentrations of Co, Mn, and Ce were 11, 2.6, and 24.8 wt.%, respectively.

The BET method was used to determine the surface specific area of each sample, and the values are also reported in **Table 1**. The surface area of the CoZ sample practically did not change with the impregnation of cobalt, showing BET value of 7.5 m^2/g, close to the ZrO_2 support. On the other hand, the surface area was similar for the CoOx-CeO_2 catalysts, regardless of the impregnation or coprecipitation methods used. As it can be observed for CoMnCeOx, the addition of a small amount of manganese increased the specific surface with respect to the other samples, possibly due to the insertion of Mn in the ceria structure.

3.2 X-ray diffraction

The crystalline structures of the different Co-based catalysts and that of the ZrO_2 and CeO_2 supports were analyzed by means of XRD. **Figure 2** shows the diffraction patterns of the samples.

The CoOx-CeO_2 catalysts (CoCe-I and CoCe-P) show patterns dominated by the characteristic CeO_2 diffraction peaks, whose main signals appear at $2\theta = 28.6°$,

33.4°, 47.8°, and 56.6° (JCPDS34-0394). The peaks of Co_3O_4 at 2θ = 31.3°, 36.8°, 44.8°, 59.4°, and 65.3° (JCPDS 42-1467) do not seem to modify the structure of the CeO_2 diffraction pattern. Despite the different preparation methods that were used, only two phases can be distinguished, i.e., Co_3O_4 spinel and CeO_2 with fluorite structure.

Catalysts	Preparation method	Co (wt.%)	Mn (wt.%)	Molar ratio of Co/ Zr or Ce	BET area (m^2/g)
CoZ	wet impregnation	9.2	—	0.15	7.5
CoCe-I	wet impregnation	10.6	—	0.14	93
CoCe-P	co-precipitation	8.6	—	0.14	92
CoCeMnOx	co-precipitation	11.0	2.6	1.06	165

Table 1.
Co-based catalysts, compositions, and specific surface area.

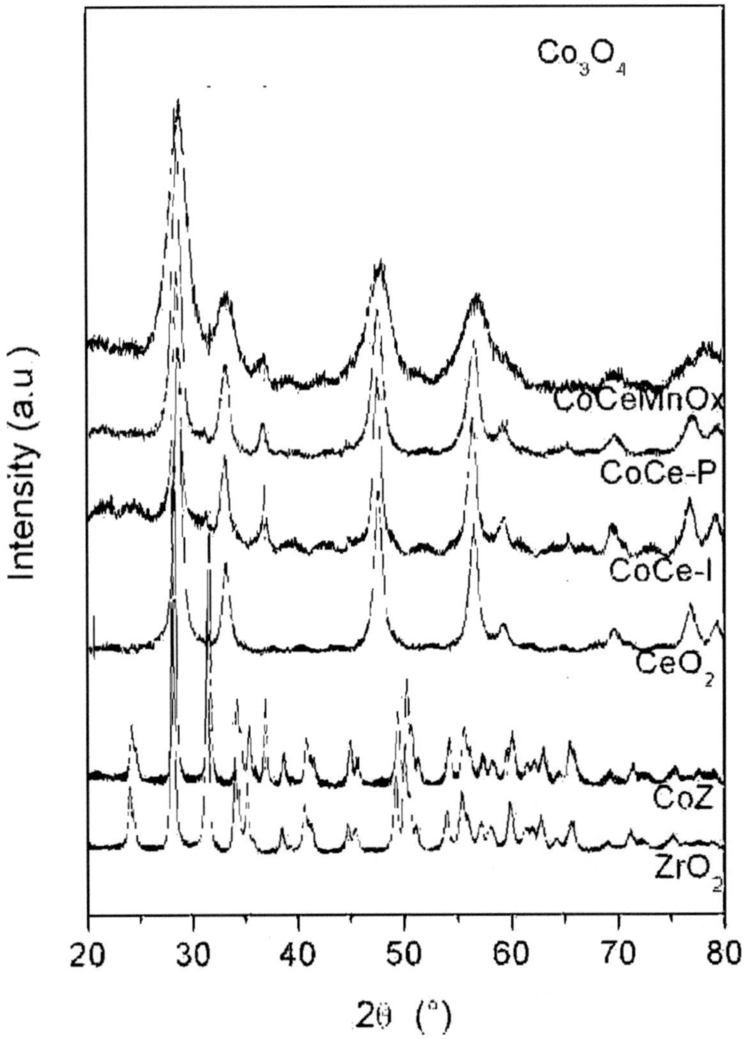

Figure 2.
XRD patterns of CoZ, CoCe-I, CoCe-P, and CoCeMnOx catalysts and CeO_2 and ZrO_2 supports.

On the other hand, the diffraction pattern of the CoZ catalyst shows the peaks associated with the monoclinic ZrO_2 support and the main signal of Co_3O_4 at $2\theta = 36.8°$.

In the diffraction pattern of the CoCeMnOx catalyst, a broadening of peaks corresponding to the CeO_2 phase can be observed. These results could indicate that the structure of CeO_2 is distorted due to the insertion of manganese, although the type fluorite structure of CeO_2 is still maintained [28]. It can be observed that in the main signal of CeO_2 ($2\theta = 28.6°$), an asymmetric broadening appears associated with an incipient formation of the $(Mn,Co)_3O_4$ phase, which has a characteristic peak at 29.4° (JCPDS 18-408). Likewise, the main peak of the $(Mn,Co)_3O_4$ phase ($2\theta = 36.4$) appears at 36.9°, close to the main signal of the Co_3O_4 spinel.

3.3 Temperature-programmed reduction

Figure 3 exhibits the reduction peaks corresponding to cobalt-based catalyst. The reduction profile of the CoZ catalyst presents a peak, whose maximum is at 395°C, and a shoulder, at 370°C. The cobalt oxide reduction is completed at 440°C. It is known that the reduction of Co_3O_4 takes place in two steps that correspond to the subsequent reduction of the Co^{3+} and Co^{2+} species present in the spinel structure.

The CoCe-P profile presents two intense peaks at 300 and 380°C belonging to Co_3O_4 reduction. The small signals at 471 and 850°C could be attributed to the reduction of the ceria support, superficial oxygen, and bulk species, respectively [29, 30]. On the other hand, in the CoCe-I sample, the peak at 471°C is a small shoulder. It is probable that for the coprecipitated catalysts, a more intimate contact between Co and Ce species takes place. This fact produces a synergistic effect promoting the reducibility of the species and the mobility of surface oxygen species in ceria support. The performance of Co_3O_4 phase on COPrOx reaction follows a redox mechanism which involves the reduction/oxidation of the Co^{3+} site. The presence of ceria and the synergism with cobalt would improve its reducibility. According to the H_2 consumption of each sample, the H_2/Co ratio close to 1.33 revealed that Co_3O_4 is the cobalt phase present on CoZ catalysts. In the case of CoOx-CeO_2 catalysts, the H_2/Co ratio values higher than 1.33 were calculated. This would indicate that a CeO_2 fraction was reduced within the range of temperatures studied.

The profile of the CoCeMnOx catalyst shows three reduction zones at low (150–420°C), intermediate (420–620°C), and high (650–900°C) temperatures. As shown by XRD, the Co species seems preferentially to form the Co_3O_4 spinel ($2\theta = 36.9°$) without distortion of the CeO_2 structure, which is reduced at low temperature. The reduction of manganese species should also be considered in this zone, because there are multiple overlapping peaks [28]. Then, the cobalt species (Co_3O_4) and a fraction of the Mn^{4+}/Mn^{3+} species incorporated to the ceria structure might be reduced below 420°C. The reduction temperatures of Co_3O_4 in the CoCeMnOx catalyst are lower than in the CoCe-I and CoCe-P samples. The other broad peak centered at 800°C is a typical reduction peak of bulk CeO_2 and could also be associated with some difficult-to-reduce Mn^{n+} species.

3.4 Laser-Raman spectroscopy

The laser-Raman spectra of catalysts and CeO_2 and ZrO_2 supports are shown in **Figure 4**. The CeO_2 spectrum shows a main band at 464 cm^{-1} (F2g mode) and other weaker signals at 259, 587, and 1173 cm^{-1} belonging to second-order (2 TA, D, and 2TO) Raman modes, respectively, which are associated to oxygen vacancies [31–33].

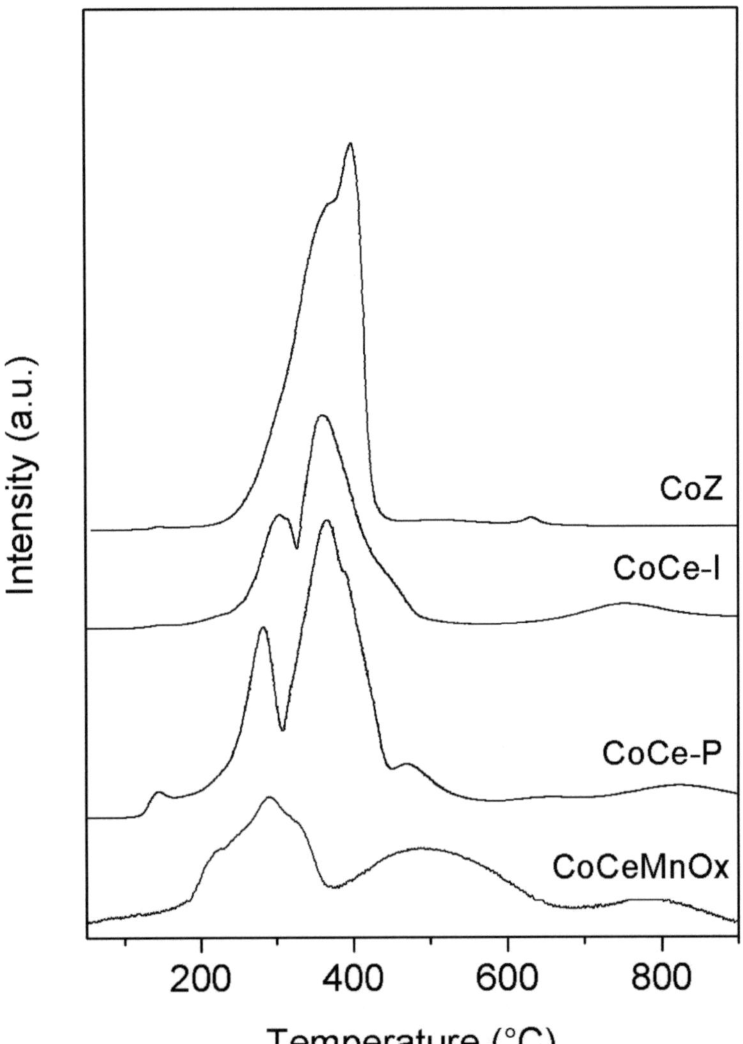

Figure 3.
Temperature-programmed reduction of CoZ, CoCe-I, CoCe-P, and CoCeMnOx catalysts.

The Raman spectra of CoCe-I and CoCe-P show a single peak at 464 cm^{-1}, which corresponds to CeO_2, as well as the signals at 193, 484, 524, 615, and 680 cm^{-1} assigned to the vibrations of the Co_3O_4 spinel. The bands at 193, 524, and 615 cm^{-1} were attributed to the F2g mode, while the bands at 484 and 680 cm^{-1} were assigned to the Eg and A1g modes, respectively [34].

In the CoCeMnOx spectrum, the bands of Co_3O_4 oxide are not detected. However, a small peak at 666 cm^{-1} is observed, due to the possible formation of a Mn,Co mixed spinel. Nevertheless, the XRD patterns showed the main peak corresponding to the Co_3O_4 spinel Therefore, the co-existence of Co_3O_4 and $(Mn,Co)_3O_4$ phases could also be possible.

The Raman spectra of the CoZ catalysts and ZrO_2 support are shown in **Figure 4**. The main bands at 680 and 520 cm^{-1} can be observed corresponding to cobalt oxides [35]. The other bands at 193, 475, and 615 cm^{-1} overlap with the bands of zirconia monoclinic phase.

In order to understand the change of the active phase under different atmospheres, the CoCe-I catalyst was subjected to different reductive/oxidative

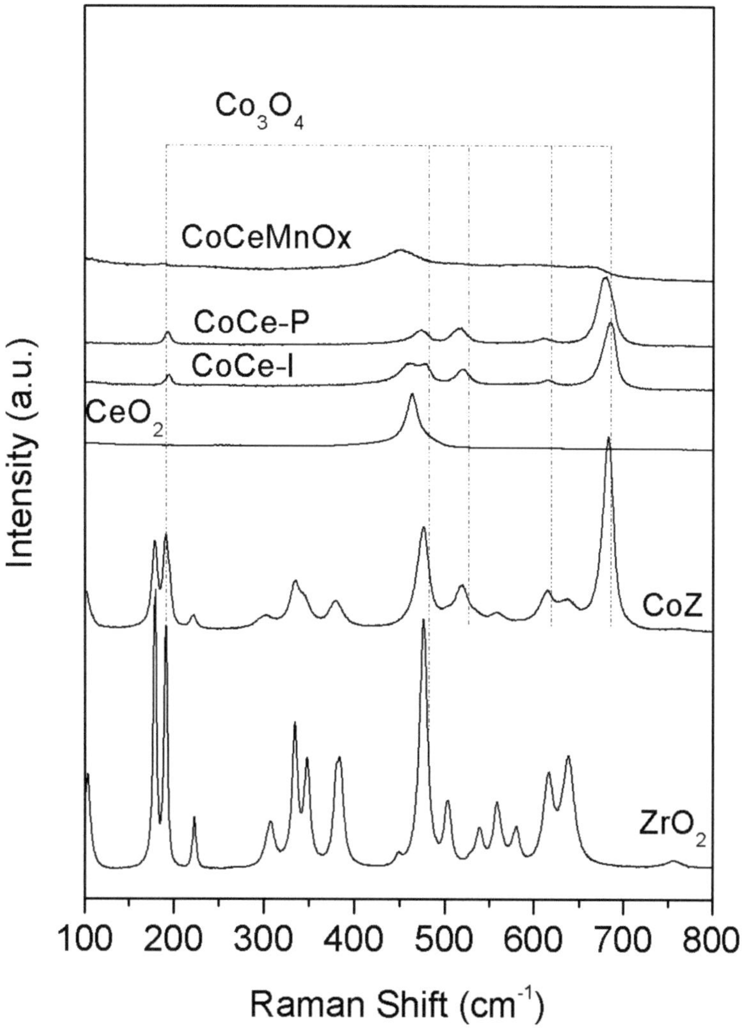

Figure 4.
Raman spectra of CoZ, CoCe-I, CoCe-P, and CoCeMnOx catalysts.

treatments at different temperatures. These experiments were followed by in situ Raman spectroscopy, recording spectra at each temperature. The catalyst was placed in a reaction chamber coupled to the spectrometer.

The first experiment consisted in reducing the catalysts with a 1% CO/Ar stream with the purpose of analyzing the CO reducing capacity. At room temperature the main signals corresponding to Co_3O_4 at 691, 521, and 484 cm^{-1}, as well as the main signal of CeO_2 at 463 cm^{-1}, were detected (**Figure 5A**). As the temperature increased, the intensity of the Co_3O_4 bands decreased. However, a weak signal around 690 cm^{-1} was distinguished at 300°C, while at 325°C no signals of Co_3O_4 were detected, which implied that Co_3O_4 was completely reduced by CO from the gas phase. The catalyst was cooled to room temperature in 1% CO/Ar mixture and then maintained in Ar atmosphere.

The following experiment consisted in oxidizing the reduced catalyst with a 1% O_2/Ar stream, and the spectra are exhibited in **Figure 5B**. In the spectrum at the bottom, recorded at room temperature, it is possible to observe the main signal of CeO_2 at 463 cm^{-1}, as well as a broad signal between 500 and 600 cm^{-1}, which was assigned to oxygen vacancies. As the temperature increases, this broad band decreases due to the incorporation of oxygen from the gas phase into the structure. Moreover, at 200°C the signals corresponding to Co_3O_4 are observable due to the oxidation of Co metal particles. After the experiment, the catalyst was cooled in 1% O_2/Ar flow at room temperature, and then it was treated in an Ar stream during several minutes. The corresponding spectrum registered at room temperature is shown in **Figure 5C**.

After that, the catalyst was again reduced at 350°C by a reducing stream (1% CO, 40% H_2 in Ar), and then, it was cooled in the same stream. A new spectrum was recorded at room temperature, and it is only possible to observe the main signal of CeO_2 at 464 cm^{-1}. Then, the catalyst was treated in Ar stream during several minutes, and oxygen vacancies were again detected.

Subsequently, the catalyst was subjected to a stream composed of the COPrOx reactant mixed gas flow (1% CO, 1% O_2, 40% H_2 and Ar). At room temperature the CeO_2 main signal was still observed. The temperature was increased, and at 150°C it was possible to detect a little band at 680 cm^{-1}, which belongs to Co_3O_4, which indicates that the oxygen present in the reactant gas stream is capable of reoxidizing the reduced cobalt species. When the temperature continued increasing, the band at 680 cm^{-1} was well defined, but at 300°C, it started again to decrease. At 325°C, the signal completely disappeared.

3.5 X-ray photoelectron spectroscopy

The surface characterization of Co-based catalysts was analyzed by the XPS technique, and the spectra for Co 2p and O 1s core levels are presented in **Figure 6**. The Co 2p spectrum shows two peaks at 780.1 and 795.0 eV, corresponding to the Co $2p_{3/2}$-Co $2p_{1/2}$ spin-orbit splitting (**Figure 6A**). These binding energies (BE) could be assigned to Co^{2+} or Co^{3+} species, since the BE difference is negligible for both species. However, the presence of intense satellite peaks is distinctive of the Co^{2+} spectrum [36, 37]. The cobalt spectra present very small satellite peaks, or else, the absence of them would indicate that Co^{3+} species are dominant for all catalysts.

Figure 6B shows the O 1s XP spectra for all of the catalysts studied. Broad and complex shape spectra are observed in this region. These spectra can be described as the result of the overlapping of two components corresponding to different

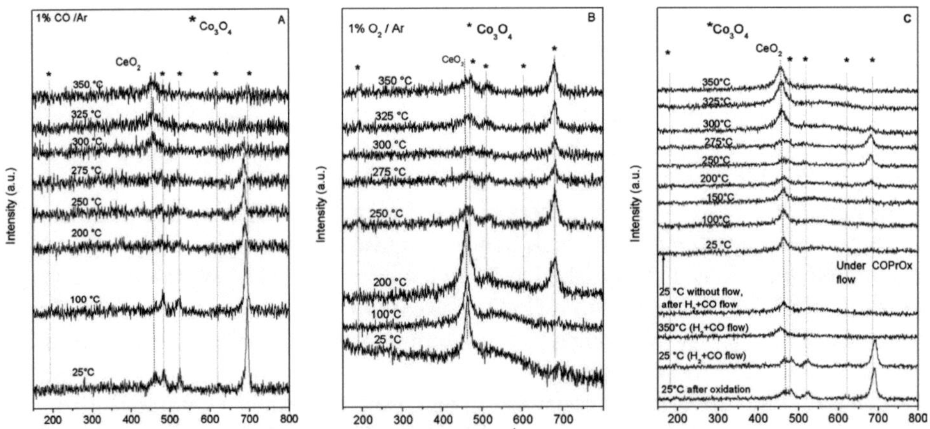

Figure 5.
In situ Raman study of CoCe-I. (A) 1% CO/Ar flow, (B) 1% O_2/Ar flow, and (C) under reaction conditions (1%CO, 1%O_2, 40% H_2, Ar).

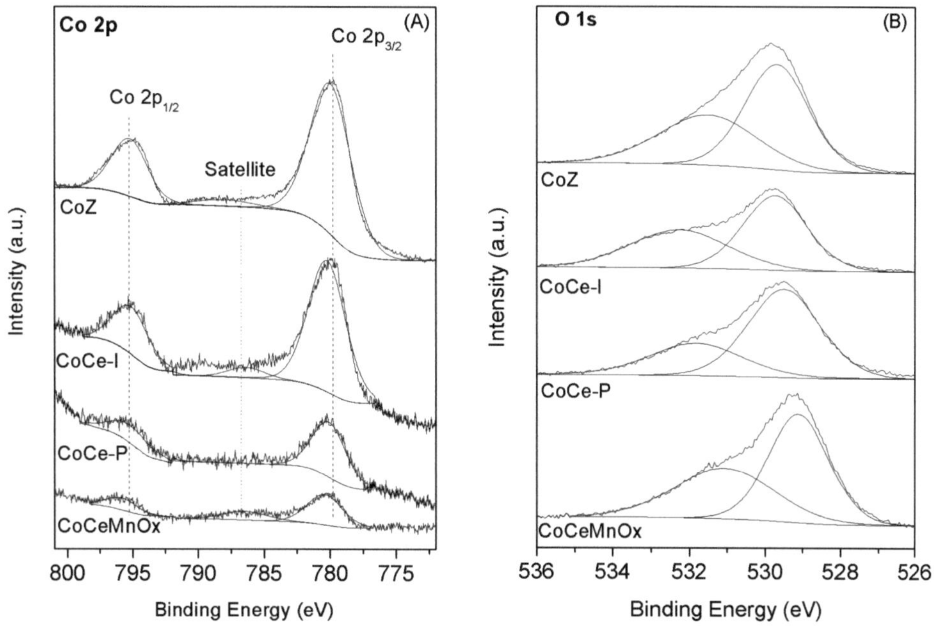

Figure 6.
XPS spectra of Co $2p_{3/2}$ (A) and O 1s (B) for CoZ, CoCe-I, CoCe-P, and CoCeMnOx catalysts.

species or groups of species. The first peak appears at 529.8 eV and the other at 532.3 eV. The oxygen signal at lower BE corresponds to bulk CeO_2 oxide and to cobalt oxides [38, 39]. The other component is assigned to hydroxyl, water, and/or carbonate groups adsorbed on the solid surface [40].

3.6 Catalytic evaluation

The COPrOx reaction system can be described by the following two reactions; the first one corresponds to CO oxidation, while the second one is the undesirable H_2 oxidation reaction.

$$CO + 0.5\,O_2 \rightarrow CO_2 \quad (5)$$

$$H_2 + 0.5\,O_2 \rightarrow H_2O \quad (6)$$

Figure 7A presents CO conversion curves of different catalysts as a function of temperature. It can be observed that the catalyst which was prepared with ZrO_2 support reached a conversion maximum (92%) lower than that reached by the catalysts prepared with cerium and at a higher temperature (200°C). CoCe-I and CoCe-P reached almost total CO conversion at 175 and 165°C, respectively, showing that CeO_2 is a better catalytic support than ZrO_2. Moreover, the small difference in catalytic performance which exists between CoCe-I and CoCe-P indicates that the coprecipitation method is slightly suitable to obtain an adequate catalyst for this reaction. When manganese was incorporated to the catalytic formulation, it could be observed that with the CoCeMnOx catalyst, complete CO conversion was reached at lower temperature than other catalysts. In addition, the temperature window of CO maxima conversion was maintained on a wide temperature range (150–200°C). **Figure 7B** shows the O_2 to CO_2 selectivity curves as a function of temperature. Selectivity curves exhibited a decreasing trend as the temperature

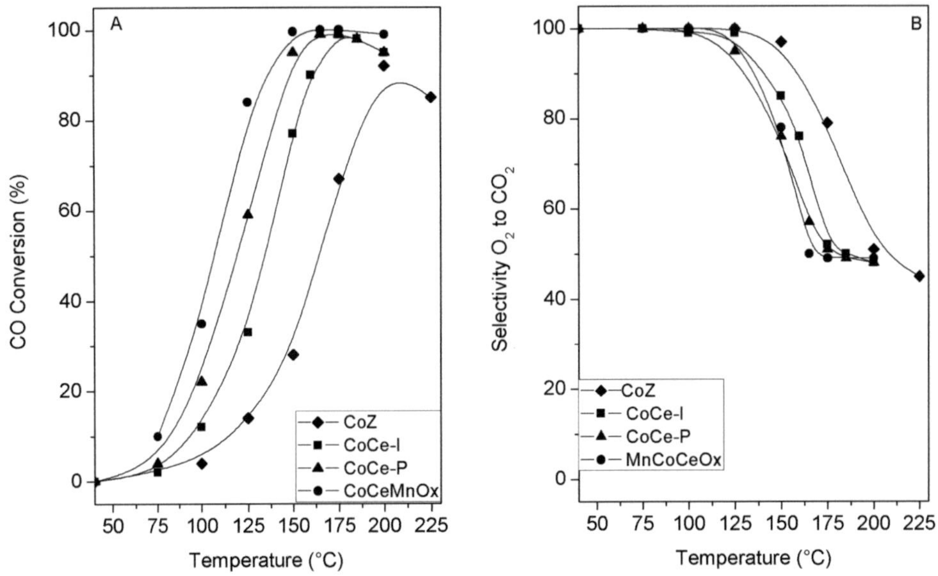

Figure 7.
COPrOx reaction on CoZ, CoCe-I, CoCe-P, and CoCeMnOx catalysts. (A) CO conversion (%). (B) O_2 selectivity to CO_2 (%). Reaction conditions: 1% CO, 1% O_2, 40% H_2 and He balance. W/F: 2.1 mg min/mL.

increased. Since the H_2 oxidation (reaction 2) had a higher apparent activation energy than the CO oxidation (reaction 1), as temperature increased, the hydrogen oxidation rate also increased so that CO conversion decreased and, consequently, the selectivity curve also decreased.

Moreover, the stability of CoCeMnOx and the effect of CO_2 adding in the feed were studied at 175°C during 75 h under reaction conditions. The CO conversion was maintained at 95% while the selectivity increased from 50 to 60%. The used catalysts did not show carbon formation by LRS spectroscopy.

The CeO_2 support with high mobility of surface lattice oxygen promotes the surface exchange between reduced and oxidized species $Ce^{4+} + Co^{2+} \leftrightarrow Ce^{3+} + Co^{3+}$. This process allows that the Co^{3+} species are available on the catalytic surface.

In the case of CoCeMnOx, which is the most active catalyst, the Co^{2+} species are oxidized by oxygen from Mn oxides or CeO_2 support, and then they are oxidized by oxygen from the gas phase. The redox couples Co^{2+}/Co^{3+} and Ce^{3+}/Ce^{4+} are improved by the presence of MnOx.

The in situ Raman results suggest that the active phase participates in the chemical reaction by a redox mechanism, where Co^{3+} oxidizes CO and reduces itself to Co^{2+}, but this cobalt species is permanently reoxidized by oxygen provided by CeO_2 and/or the gas phase. These experiments also allow inferring that the reduction step limits the overall reaction while the reoxidation on the surface is very fast.

4. Conclusions

The nature of cobalt species observed on Co-based catalysts used on the COPrOx reaction was influenced by the constituent species of the environment.

The XRD, TPR, XPS, and LRS characterizations showed that the Co_3O_4 active phase was present in all catalysts. From diffraction patterns of CoOx-ZrO_2 and the CoOx-CeO_2 catalysts, the Co_3O_4 spinel and the signals corresponding to ZrO_2 and CeO_2 were identified in agreement with laser-Raman spectra.

The CoCeMnOx catalyst, prepared by coprecipitation of precursor salts, showed an incipient development of a new phase $(Mn,Co)_3O_4$ mixed spinel, due to the intimate contact between elements.

The reduction of cobalt oxides was completed below 400°C and was promoted by the oxygen vacancies of ceria. In contrast, in the CoZ catalysts, the cobalt species were totally reduced at 440°C. The profiles showed two reduction steps assigned to Co^{3+} to Co^{2+} and then Co^{2+} to Co^{0}.

In situ laser-Raman spectroscopy revealed that cobalt species remained oxidized in spite of the reducing atmosphere of the reactant flow within the temperature range studied.

The XPS spectra showed the coexistence of Co^{2+} and Co^{3+} species according to the Co_3O_4 phase on the catalytic surface.

The CO conversion reached by the $CoOx\text{-}CeO_2$ catalysts was higher than the $CoOx\text{-}ZrO_2$ sample, due to redox properties of ceria. The manganese addition promoted the catalytic activity, widening the temperature window of maximum CO conversion.

Acknowledgements

The authors acknowledge the financial support received from UNL, ANPCyT, and CONICET. Thanks are given to Fernanda Mori for the XPS measurements.

Conflict of interest

The authors declare no conflict of interest with any person, institution, or organization for the publication of this chapter.

Author details

Leticia E. Gómez and Alicia V. Boix*
School of Chemical Engineering, Institute of Research on Catalysis and Petrochemistry (INCAPE), Universidad Nacional del Litoral (FIQ-UNL), Santa Fe, Argentina

*Address all correspondence to: aboix@fiq.unl.edu.ar

References

[1] Abbott D. Keeping the energy debate clean: How do we supply the world's energy needs? Proceedings of the IEEE. 2010;**98**:42-66. DOI: 10.1109/JPROC.2009.2035162

[2] Baykara SZ. Hydrogen: A brief overview on its sources, production and environmental impact. International Journal of Hydrogen Energy. 2018;**43**:10605-10614. DOI: 10.1016/j.ijhydene.2018.02.022

[3] Brandon NP, Kurban Z. Clean energy and the hydrogen economy. Physical and Engineering Sciences. 2017;**375**:20160400. DOI: 10.1098/rsta.2016.0400

[4] Wang Y, Chen KS, Mishler J, Cho SC, Adroher XC. A review of polymer electrolyte membrane fuel cells: Technology, applications, and needs on fundamental research. Applied Energy. 2010;**88**:981-1007. DOI: 10.1016/j.apenergy.2010.09.030

[5] Bion N, Epron F, Moreno M, Mariño F, Duprez D. Preferential oxidation of carbon monoxide in the presence of hydrogen (PROX) over noble metals and transition metal oxides: Advantages and drawbacks. Topics in Catalysis. 2008;**51**:76-88. DOI: 10.1007/s11244-008-9116-x

[6] Park ED, Lee D, Lee HC. Recent progress in selective CO removal in a H_2-rich stream. Catalysis Today. 2009;**139**:280-290. DOI: 10.1016/j.cattod.2008.06.027

[7] Gómez LE, Sollier BM, Mizrahi M, Ramallo López JM, Miró EE, Boix AV. Preferential CO oxidation on Pt-Cu/Al_2O_3 catalysts with low Pt loadings. International Journal of Hydrogen Energy. 2014;**39**:3719-3729. DOI: 10.1016/j.ijhydene.2013.12.146

[8] Serra RM, Aspromonte SG, Miró EE, Boix AV. Hydrocarbon adsorption and NOx-SCR on (Cs,Co) mordenite. Applied Catalysis B: Environmental. 2015;**166-167**:592-602. DOI: 10.1016/j.apcatb.2014.11.061

[9] Banús ED, Milt VG, Miró EE, Ulla MA. Catalytic coating synthesized onto cordierite monolith walls. Its application to diesel soot combustión. Applied Catalysis B: Environmental. 2013;**132-133**:479-486. DOI: 10.1016/j.apcatb.2012.12.020

[10] Coronel L, Múnera JF, Tarditi AM, Moreno MS, Cornaglia LM. Hydrogen production by ethanol steam reforming over Rh nanoparticles supported on lanthana/silica systems. Applied Catalysis B: Environmental. 2014;**160-161**:254-266. DOI: 10.1016/j.apcatb.2014.05.025

[11] Dalla Fontana A, Faroldi B, Cornaglia LM, Tarditi A. Development of catalytic membranes over PdAu selective films for hydrogen production through the dry reforming of methane. Molecular Catalysis. DOI: 10.1016/j.mcat.2018.07.018

[12] Gómez LE, Tiscornia IS, Boix AV, Miró EE. Co/ZrO_2 catalysts coated on cordierite monoliths for CO preferential oxidation. Applied Catalysis A: General. 2011;**401**:124-133. DOI: 10.1016/j.apcata.2011.05.007

[13] Gómez LE, Tiscornia IS, Boix AV, Miró EE. CO preferential oxidation on cordierite monoliths coated with Co/CeO_2 catalysts. International Journal of Hydrogen Energy. 2012;**37**:14812-14819. DOI: 10.1016/j.ijhydene.2012.01.159

[14] Banús ED, Ulla MA, Miró EE, Milt VG. Co,Ba,K/ZrO_2 coated onto metallic foam (AISI 314) as a structured catalyst for soot combustion: Catalytic activity and stability. Applied Catalysis A: General. 2011;**393**:9-16. DOI: 10.1016/j.apcata.2010.11.018

[15] Boix AV, Aspromonte SG, Miró EE. Deactivation studies of the SCR of NOx with hydrocarbons on Co-mordenite monolithic catalysts. Applied Catalysis A: General. 2008;**341**:26-34. DOI: 10.1016/j.apcata.2007.12.032

[16] Gómez LE, Boix AV, Miró EE. Co/ZrO_2, Co/CeO_2 and MnCoCe structured catalysts for COPrOx. Catalysis Today. 2013;**216**:246-253. DOI: 10.1016/j.cattod.2013.05.010

[17] Liu K, Wang A, Zhang T. Recent advances in preferential oxidation of CO reaction over platinum group metal catalysts. ACS Catalysis. 2012;**2**:1165-1178. DOI: 10.1021/cs200418w

[18] Komatsu T, Takasaki M, Ozawa K, Furukawa S, Muramatsu A. PtCu intermetallic compound supported on alumina active for preferential oxidation of CO in hydrogen. Journal of Physical Chemistry C. 2013;**117**:10483-10491. DOI: 10.1021/jp4007729

[19] Korotkikh O, Farrauto R. Selective catalytic oxidation of CO in H_2: Fuel cell applications. Catalysis Today. 2000;**62**:249-254. DOI: 10.1016/S0920-5861(00)00426-0

[20] Marbán G, Fuertes AB. Highly active and selective CuOx/CeO_2 catalyst prepared by a single-step citrate method for preferential oxidation of carbon monoxide. Applied Catalysis B: Environmental. 2005;**57**:43-53. DOI: 10.1016/j.apcatb.2004.10.011

[21] Jampa S, Wangkawee K, Tantisriyanurak S, Changpradit J, Jamieson AM, Chaisuwan T, et al. High performance and stability of copper loading on mesoporous ceria catalyst for preferential oxidation of CO in presence of excess of hydrogen. International Journal of Hydrogen Energy. 2017;**42**:5537-5548. DOI: 10.1016/j.ijhydene.2016.08.078

[22] Lacoste AM, Tiscornia IS, Boix AV. CO preferential oxidation on cordierite monoliths coated with CuO-CeO_2/SBA-15 catalysts. Further insights into the physico-chemical aspects of the catalytic behavior. International Journal of Hydrogen Energy. 2018;**43**:14238-14251. DOI: 10.1016/j.ijhydene.2018.05.122

[23] Liotta LF, Di Carlo G, Pantaleo G, Venezia AM, Deganello G. Co_3O_4/CeO_2 composite oxides for methane emissions abatement: Relationship between Co_3O_4-CeO_2 interaction and catalytic activity. Applied Catalysis B: Environmental. 2006;**66**:217-227. DOI: 10.1016/j.apcatb.2006.03.018

[24] Gómez LE, Múnera JF, Sollier BM, Miró EE, Boix AV. Raman in situ characterization of the species present in Co/CeO_2 and Co/ZrO_2 catalysts during the COPrOx reaction. International Journal of Hydrogen Energy. 2016;**41**:4993-5002. DOI: 10.1016/j.ijhydene.2016.01.099

[25] Zhao Z, Jin R, Bao T, Yang H, Lin X, Wang G. Mesoporous $Ce_xMn_{1-x}O_2$ composites as novel alternative carriers of supported Co_3O_4 catalysts for CO preferential oxidation in H_2 stream. International Journal of Hydrogen Energy. 2012;**37**:4774-4786. DOI: 10.1016/j.ijhydene.2011.12.057

[26] Zhao Z, Lin X, Jin R, Wang G, Muhammad T. MOx (M=Mn, Fe, Ni or Cr) improved supported Co_3O_4 catalysts on ceria-zirconia nanoparticulate for CO preferential oxidation in H_2-rich gases. Applied Catalysis B: Environmental. 2012;**115-116**:53-62. DOI: 10.1016/j.apcatb.2011.12.001

[27] Gómez LE, Miró EE, Boix AV. Spectroscopic characterization of Mn-Co-Ce mixed oxides, active catalysts for COPROX reaction. International Journal of Hydrogen Energy. 2013;**38**:5645-5654. DOI: 10.1016/j.ijhydene.2013.03.004

[28] Picasso G, Gutiérrez M, Pina MP, Herguido J. Preparation and characterization of Ce-Zr and Ce-Mn based oxides for n-hexane combustion: Application to catalytic membrane reactors. Chemical Engineering Journal. 2007;**126**:119-130. DOI: 10.1016/j.cej.2006.09.005

[29] Trovarelli A. Catalytic properties of ceria and CeO_2-containing materials. Catalysis Reviews: Science and Engineering. 1996;**38**(4):439-520. DOI: 10.1080/01614949608006464

[30] Luo JY, Meng M, Li X, Li XG, Zha YQ, Hu TD, et al. Mesoporous Co_3O_4-CeO_2 and Pd/Co_3O_4-CeO_2 catalysts: Synthesis, characterization and mechanistic study of their catalytic properties for low-temperature CO oxidation. Journal of Catalysis. 2008;**254**:310-324. DOI: 10.1016/j.jcat.2008.01.007

[31] Wu Z, Li M, Howe J, Meyer HM, Overbury SH. Probing defect sites on CeO_2 nanocrystals with well-defined surface planes by Raman spectroscopy and O_2 adsorption. Langmuir. 2010;**26**:16595-16606. DOI: 10.1021/la101723w

[32] Zdravković J, Simović B, Golubović A, Poleti D, Veljković I, Šćepanović M, et al. Comparative study of CeO_2 nanopowders obtained by the hydrothermal method from various precursors. Ceramics International. 2015;**41**:1970-1979. DOI: 10.1016/j.ceramint.2014.08.122

[33] Lee Y, He G, Akey A, Si R, Flytzani-Stephanopoulos M, Herman IP. Raman analysis of mode softening in nanoparticle $CeO_{2-\delta}$ and Au-$CeO_{2-\delta}$ during CO oxidation. Journal of the American Chemical Society. 2011;**133**:12952-12955. DOI: 10.1021/ja204479j

[34] Woods MP, Gawade P, Tan B, Ozkan US. Preferential oxidation of carbon monoxide on Co/CeO_2 nanoparticles. Applied Catalysis B: Environmental. 2010;**97**:28-35. DOI: 10.1016/j.apcatb.2010.03.015

[35] Tang CW, Wang CB, Chien SH. Characterization of cobalt oxides studied by FT-IR, Raman, TPR and TG-MS. Thermochimica Acta. 2008;**473**:68-73. DOI: 10.1016/j.tca.2008.04.015

[36] Hagelin-Weaver HAE, Hoflund GB, Minahan DM, Salaita GN. Electron energy loss spectroscopic investigation of Co metal, CoO, and Co_3O_4 before and after Ar^+ bombardment. Applied Surface Science. 2004;**235**:420-448. DOI: 10.1016/j.apsusc.2004.02.062

[37] Pietrogiacomi D, Tuti S, Campa MA, Indovina V. Cobalt supported on ZrO_2: Catalysts characterization and their activity for the reduction of NO with C_3H_6 in the presence of excess O_2. Applied Catalysis B: Environmental. 2000;**28**:43-54. DOI: 10.1016/S0926-3373(00)00161-2

[38] Liu J, Zhao Z, Wang J, Xu C, Duan A, Jiang G, et al. The highly active catalysts of nanometric CeO_2-supported cobalt oxides for soot combustion. Applied Catalysis B: Environmental. 2008;**84**:185-195. DOI: 10.1016/j.apcatb.2008.03.017

[39] Galtayries A, Sporken R, Riga J, Blanchard G, Caudano R. XPS comparative study of ceria/zirconia mixed oxides: Powders and thin film characterization. Journal of Electron Spectroscopy and Related Phenomena. 1998;**88-91**:951-956. DOI: 10.1016/S0368-2048(97)00134-5

[40] Alvarez M, López T, Odriozola JA, Centeno MA, Domínguez MI, Montes M, et al. 2,4-Dichlorophenoxyacetic acid (2,4-D) photodegradation using an Mn^+/ZrO_2photocatalyst: XPS, UV-vis, XRD characterization. Applied Catalysis B: Environmental. 2007;**73**:34-41. DOI: 10.1016/j.apcatb.2006.12.010

Section 5

Cobalt Single Atom Heterogeneous Catalyst: Method of Preparation, Characterization, Catalysis and Mechanism

Chapter 5

Cobalt Single Atom Heterogeneous Catalyst: Method of Preparation, Characterization, Catalysis, and Mechanism

Baljeet Singh, Surender Kumar and Archana Singh

Abstract

Transition metal nanoparticles and metal oxide have been used extensively for a wide range of applications in electrochemical reactions (HER, ORR, OER) and energy storage (supercapacitors batteries). To make less expensive, the use of transition metal at minimum metal contents without compromising the catalytic activity could be one way. Most of the catalytic process takes place on the surface and reaction dynamic can be manipulated by changing the particle size and shape. For a long time, single metal atom organometallic compounds have been used as a catalyst at the industrial level. However, problems with the homogeneous catalyst to recover back at the end of the process lead to development of heterogeneous single-atom catalysts with equal activity like a homogeneous catalyst. Cobalt single atom has received a tremendous interest of the scientific community due to its excellent catalytic activity and recyclability. Cobalt single-atom catalyst has shown better performance compared with sub-nanometer nanoparticles catalyst for ORR, OER, and HER. This chapter is conferring method of preparation of carbon-based single Co atoms heterogeneous catalyst, their application for ORR, OER, HER reactions, and mechanistic investigations through DFT calculations. The role of single Co metal atoms and anchoring using *N* or heteroatoms is discussed and compared.

Keywords: single metal atom heterogeneous catalyst, transition metal catalyst, homogeneous, heterogeneous, oxygen reduction reaction (ORR), oxygen evolution reaction (OER) and hydrogen evolution reaction (HER)

1. Introduction

The main aim of current research on catalysis is to find out long-lasting catalyst that can consistently work without compromising its catalytic activity [1–3]. The catalytic activity mostly depends on a number of active sites and their accessibilities. Catalytic activity can be tuned by manipulating the nanoparticles sizes or by producing shaped nanoparticles with exposed facets [2–5]. Keys factors for catalytic properties are not very well known and atoms with the different environments are mostly showed different catalytic activity. For example, an atom on the edges and on corners mostly showed different properties, more catalytic activity compared with the atoms present in the middle of exposed facet.

For catalysis purpose, mostly surface atoms are accessible and catalytically active, remaining bulk atoms are providing support. The amount of metal required to produce catalytic activity can be reduced by bringing down the particle size at the atomic level, it may be enhanced by using single-site metal catalyst. Single-metal atom catalysts (SMACs) have shown better activity compared with sub-nanometer nanoparticles materials. As limited by the change in morphologies of nanoparticles, a concept of SMACs has emerged because it only way remaining to maximize efficiency of catalyst [4, 5]. Many single Co atoms homogeneous catalysts have been reported (**Figure 1a, b**) [7], in which single metal atom is supported by the bulky organic functional groups; their activity and accessibility depends on the organic environment around the single metal atom. These organic functional groups do not allow to agglomerate, provide thermodynamic stability and chemical stability (**Figure 1**). However, it is very tedious and laborious to recover back homogeneous catalyst from solution mixture and hard to reuse it again. Thus we need a catalytic system that is easy to recover and reusable. So, heterogeneous catalysts can be a good choice. However, to achieve the catalytic efficiency equal to homogeneous catalyst, a single-atom catalyst preparation is still challenging.

The preparation of stable SMACs is still challenging because high energy of a single atom leads to agglomeration and makes them unstable under reaction conditions at the industrial level. An effective way for a synthesis of SMACs via increasing the interaction between a metal atom and support [4]. In general process, by reducing the size of nanocatalyst and bring them at atomic scale, we are trying to heterogenize homogeneous catalysts on heterogeneous supports. This can be done by overcoming surface energy of a single atom by anchoring on a support substrate, which might sufficiently active like a homogenous catalyst. Graphene is a good support to stabilize coordinately unsaturated single metal sites (CUMS). Graphene is a

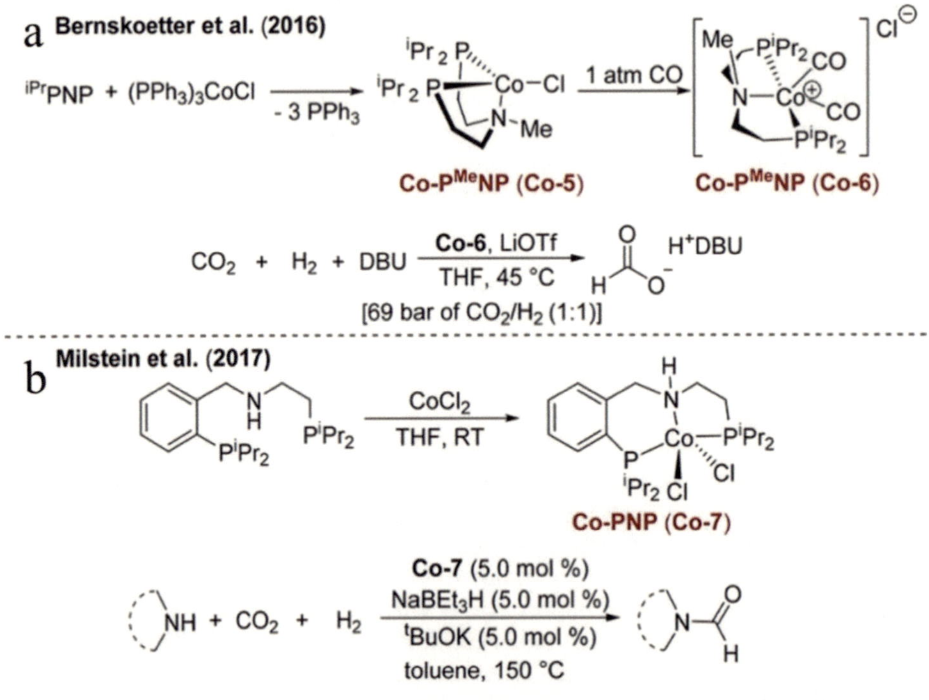

Figure 1.
(a and b) Co-PNP complex catalyzed the hydrogenation of CO2 to formate and formamides (Reproduced with the permission, Copyright 2018, American Chemical Society [7]).

perfect two-dimensional structure with high specific surface area, high mechanical strength, and thermal stability. Its unique structure and electronic properties are promising for the synthesis of stable CUMS metal atom. Several recent works have been demonstrated the preparation of SMACs using graphene [6–11].

Overall support substrate and metal interaction could help to design long-lasting, stable SMACs which could work without loss of catalytic activity under the operational condition for specific reaction or applications. Using support-metal interaction concept, Sahoo *et al*. investigated a theoretical model of graphene supported transition metal (TM) atom including Iron (Fe) single atom [8]. DFT result showed that graphene supported single-atom catalyst display relative lower activation energy barrier for methane activation. However, of its excellent catalytic performance, it has been proved that a single metal atom can migrate on a surface of graphene defect (**Figure 2**), which allows them to agglomerate into bigger nanoparticles, resultant in loss of catalytic activity. It has been proved that a single metal atom can migrate from one position to another position on graphene layer. Zhao *et al*. observed single Fe atom diffusion on graphene edge [12]. *In situ* electron microscopy was used to investigate diffusion of Fe along edges of graphene via either adding carbon atoms or by removing carbon atom (**Figure 2A–G**). Besley *et al.* also studied the dynamic behavior of single Fe atoms embedded in graphene sheet [13]. Migration of Fe atom was also observed by the aberration-corrected high-resolution transmission electron microscopy (AC-HRTEM) experiments. Anchoring mechanism using heteroatoms has found most effective tool

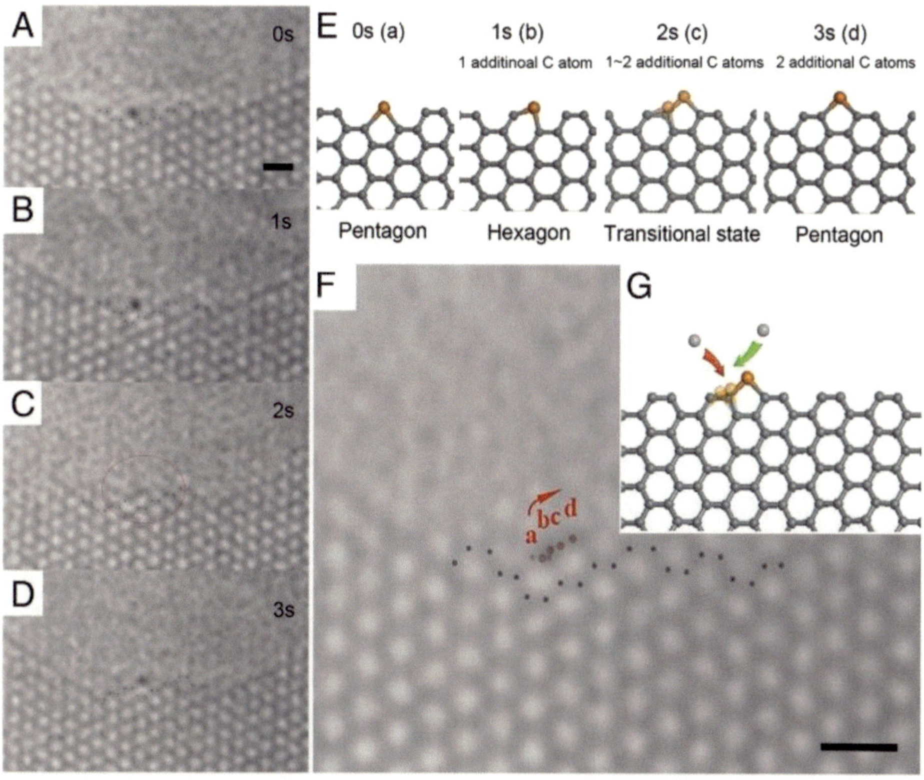

Figure 2.
One cycle growth of graphene edge: (A–D) a series of high-resolution TEM images for 4 s, Fe atom are highlighted with the red dots, nearby carbon atoms are highlighted by black dots. (E) Corresponding atomic structure of A–D. (F) The combination of A–D, which showed the trace of the Fe atom during one unit cell translocations, (G) the atomic structure of the whole growth process (Reproduced with the permission, 2014, Proceedings of the National Academy of Science of the United States of America [12]).

for synthesis of stable SMACs, which could avoid their agglomeration and helps in remain catalytically active. However, anchoring mechanisms is not known completely and this mechanism can vary with substrates. In case of pristine graphene-SMACs void or defect in the single-layer graphene served as a trap site for the single metal atom and help them to stabilize.

2. Synthesis characterization and applications of single metal atom heterogeneous catalyst

The main objective of synthesis of SMACs is 100% dispersion of metal atom on a solid support (**Figure 3**). In SMACs, metal sites may contain different numbers of metal atoms like dimer, monomer, trimer or cluster of a single atom. The main feature of SMACs is highly active low coordination metal site which attracts reactant molecules effectively, and excellent overall rate of reaction. The most important feature of SMACs is metal support interaction which affects the electronic structure and bond formation capability with other molecules in reaction media [15]. However, it is quite challenging to synthesis isolated single metal atoms on desired support. For long-lasting stable SMACs, the interaction between a metal atom and support should be strong. Otherwise, these metal atoms can leached out or migrate on surface of support and slowly convert into metal nanoparticles during reaction process. For synthesis of SMACs, many techniques have been used, including wet chemistry method, atomic layer deposition, modular synthesis based on metal–organic framework, and mass-selected soft-landing method [16]. Out of all techniques, pyrolysis and wet chemistry synthesis are mostly preferred. In pyrolysis, samples are heated at elevated temperature > 500°C under the protected atmosphere of different gases including N_2, Argon, NH_3, and H_2. For the synthesis of carbon or graphene-based SMACs, pyrolysis at high temperatures

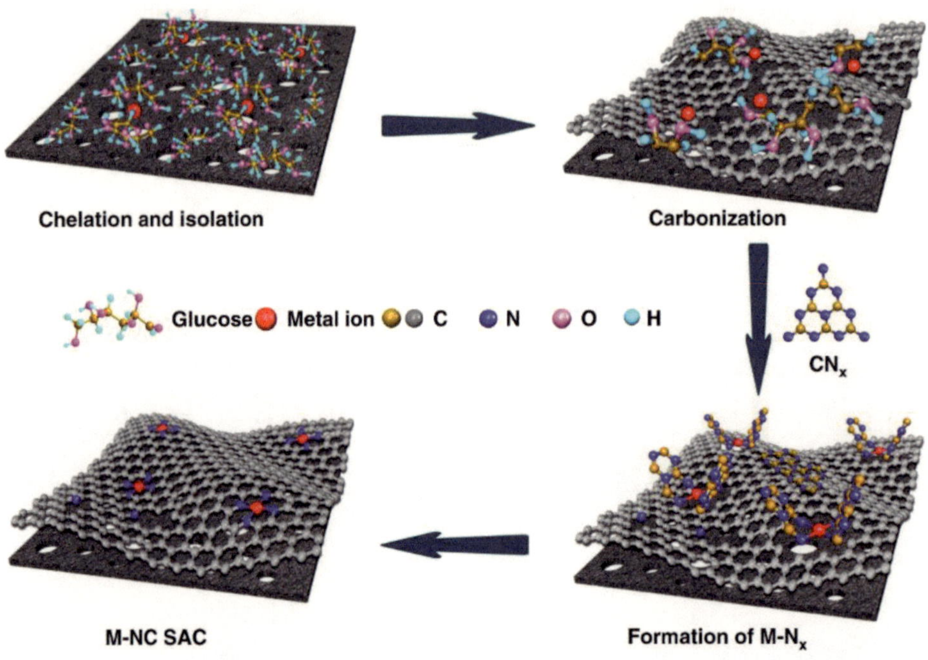

Figure 3.
Demonstration of TM single atom anchoring on a carbon support, (Reproduced with the permission, 2019, Nature Publishing Group [14]).

at inert gas conditions has been used extensively. Carbon-based materials including graphite [5], graphene [17, 18], nitrogen-doped graphene [19] were proved to be a suitable support for the synthesis of Co SMACs, because of abundant carbon vacancies at anchoring sites. In general, SMACs synthesis divided into four categories [20].

a. **Use of metal–organic framework (MOF) for SMACs synthesis:** In recent years use of MOF in heterogeneous catalyst continuously expanding because it well defines porous structure, high-ordered metal and organic arrangements, well-separated metal center, and multi-metallic. MOF is well defined single metal atom catalyst, however, due to its thermal and chemical stability significantly hamper the widespread uses. Also, as prepared MOF cannot be used for electrocatalytic applications, only few examples known in literature [21]. One way is to mitigate these issues to convert them into carbon-supported SMACs. Carbon-based materials do not only provide anchoring with the graphitic skeleton, but also provide thermal and chemical stability to the single metal atoms. Conducting graphitic two-dimensional nature of carbon support help in manipulating charge density and electronic structure of the metal atoms.

b. **Use of small metal-containing organic molecules for SMACs synthesis:** In this procedure as prepared carbon skeletons are mixed with the metal-containing small organic molecules and subsequent heating at high temperature in the control environment give a single metal atoms doped material [20]. These molecular precursors are mostly low boiling or melting point, low thermal decomposition temperature which creates difficulties during the carbonization of these precursors with the support materials. Hence for the synthesis of SMACs, the selection of molecular precursors is very crucial and important. For example, Chen *et al.* used silver tricyanomethanide and cyanamide as a starting molecular precursor [22]. Both precursors have nitrogen and cyanamide mostly used for the synthesis of g-C_3N_4. Carbonization of a mixture of both leads to the formation of C_3N_4 nanosheet doped with the silver atoms. In this procedure, silver precursors can be replaced with the porphyrin complexes of iron, zinc, or cobalt molecular precursors for the synthesis of different metal atoms on C_3N_4. Besides this, a mixture of two different molecular precursors can help in the preparation of a multi-metallic single-atom catalyst (MM-SACs).

c. **Use of sacrificial or hard template for SMACs synthesis:** During the heterogenous catalytic reaction, reactant and product undergoes a series of reactions over the catalyst surface: (1) diffusion of reactant through the solvent used to disperse catalyst, (2) diffusion of reactant in the pore of the catalyst to access catalytic sites, (3) adsorption of reactant on active site, (4) surface reaction or catalysis involved surface diffusion, (5) desorption of product or byproduct from catalytic sites, (6) diffusion of product or bi-product through pores, and (7) diffusion of the product in solvent used for catalysis [20]. Hence, surface area, pore size distribution, and accessibility of active sites are very important parameters for good catalytic activity. Surface area and pore size of catalyst can be manipulated using the templates like silica which can be sacrificed at the end of synthesis to produce a porous network. For example, Liang *et al.* synthesized a series of Co and Fe single atoms heterogeneous catalysts in a mesoporous carbon matrix using silica as a hard template [23]. After pyrolysis in a controlled environment, silica can be leached out with the help of base using HF or NaOH aqueous solutions. Besides that, metal oxide including MgO or $Mg(OH)_2$ can also be used for the synthesis of SMACs.

d. **Use of metal-containing complexes with carbon matrix for SMACs synthesis:** this procedure widely used for the synthesis of carbon-based SAMCs. Dispersed carbon nanoparticles or graphene oxide was mixed with the metal salt and other necessary chemicals [20]. Heating in a close reactor at high temperature and pressure leads to the formation of metal embedded carbon hydrogel. Dried aerogels are being heated at high temperatures in NH_3 atmosphere which helps in creating an *N* anchoring site in graphene oxide sheets or carbon networks. In final materials, the M-N_4 site formed, which can be conformed using surface characterization techniques including high-angle annular dark-field scanning transmission electron microscopy (HAADF-STEM) and Extended X-ray absorption fine structure (EXAFS). NH_3 treatment can also be used for creating an *N* anchoring site in previously synthesized porous carbon along with the metal salt mixture. This leads to the formation of single metal salt along with bigger metal nanoparticles, which can be eliminated using post-synthesis acid treatment.

2.1 Hydrogen evolution reaction (HER)

H_2 is a next-generation green fuel and H_2 generation using HER can leads to sustainable ways to produce H_2 which can be used for generation of electricity in fuel cells of combustion engine or directly can be used in H_2 engine for transportation. To achieve the high efficiency of water splitting, use of an efficient catalyst to minimize the overpotential is required. Noble metals nanoparticle including Pt, Ag, and Au are known for HER activity and requires very small overpotential with the high reaction rate in acidic solution. However of its excellent catalytic activity, due to high cost, and limited availability, we urgently need to develop an efficient transition metal (TM) catalyst. Co is a known TM, available in large amounts and also not very expensive. The development of Co single-atom catalyst could further reduce the amount of metal required to achieve catalytic efficiency like noble metals.

Water electrolysis is performed by passing a current between two electrodes through the acidic or basic solution. Overall HER reaction consists of two half-reactions, at cathode hydrogen evolution reaction (HER) where H_2 is formed and at anode, oxygen evolution reaction (OER) where O_2 is formed. These reaction mechanisms can vary using pH of the reaction media. In acidic medium, the active ions are proton and in alkaline medium, hydroxide ions are active species (below reaction). HER reaction ($2H^+ + 2e^- = H_2$) is a two-electron transfer cathodic reaction in electrochemical water splitting.

Acidic electrolyte:

$2H^+(aq) + 2e^- \rightarrow H_2(g)$
Volmer step: $H^+ + e^- \rightarrow H_{ads}$
Tafel step: $2H_{ads} \rightarrow H_2$
Heyrovsky step: $H^+ + H_{ads} + e^- \rightarrow H_2$

Alkaline electrolyte:

$2H_2O + 2e^- \rightarrow H_2\,(g) + 2OH^-$
Volmer step water dissociation: $2H_2O + 2e^- \rightarrow H_{ads}\,(g) + 2OH^-$
Tafel step: $2H_{ads} \rightarrow H_2$
Heyrovsky step: $H_2O + H_{ads} + e^- \rightarrow H_2 + OH^-$

Water molecules are the sources of proton in alkaline medium; they react with the OH^- ions, resultant in the formation of hydrogen atom adsorbed on the surface of electrode. The mechanism of HER reaction can be clearly observed using cyclic voltammetry (CV). A CV is recorded mostly in the range of maximum and minimum potential values. HER mechanism has three reaction steps including Volmer, Heyrovsky, and Tafel. This reaction can have different pathways on the cathode and mostly followed the combination of two reaction pathways. Overpotential is defined as the difference between the equilibrium between thermodynamic potential and the potential at which catalyst operates at a specific current at given operational conditions.

$$\eta = E_{applied} - E_t$$

where η is the overpotential (V), $E_{applied}$ is applied potential (V), E_t is thermodynamic potential (V). Tafel plots in electrochemistry are used to understand the kinetics of electrochemical cell reactions on the surface of catalyst. This can be used for both anodic and cathodic reactions. Final equation included two Tafel variables including Tafel slope and exchange current density.

$$\eta = \beta * log\left(\frac{i}{i_0}\right)$$

where β is the Tafel slope (mV/dec), i is the current density (A/m^2), i_0 is the exchange current density (A/m^2). β is a Tafel slope described the slope of the linear region (**Figure 4c**) within the Tafel plot. A Tafel slope gives information about the rate of reactions; which is inherent properties of the catalyst. For example, a Pt catalyst in acidic medium, Tafel slope is 120 mV for Volmer step, 40 mV for Heyrovsky step and 30 mV for Tafel step. For a good electrocatalyst, a small Tafel slope value

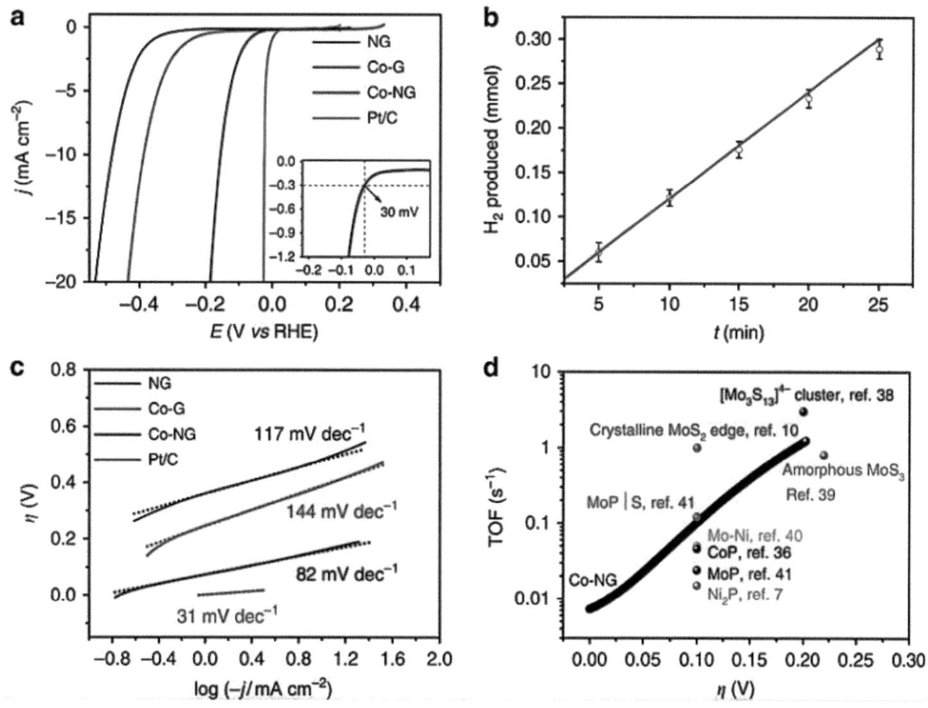

Figure 4.
HER reaction performance (a) LSV of Co-NG, NG and Co-G in 0.5 M H_2SO_4 at the scan rate of 2 mVs^{-1}. (b) Rate of H_2 production per unit time. (c) Tafel plots of the polarization curve. (d) TOF value of the Co-NG catalyst along with the TOF values of the recent reports (Reproduced with the permission, 2015, Nature Publishing Group [19]).

is expected. Exchange current density of the electrochemical reactions is another important property of the electrocatalyst. For a good catalyst, current density should be higher. Current density is closely related to reaction rate in the equilibrium state. Higher current density always more stable and does not affect by the external energy sources including temperature. Current density can be calculated using Tafel slope.

Fei *et al*. introduced an inexpensive, easy, and scalable method for synthesis of Co SACs on nitrogen-doped graphene oxide (GO) [19]. A pre-treated mixture of graphene oxide and a small amount of cobalt salt was heated in an NH_3 gas environment. For the synthesis of Co-NG, an aqueous solution of GO was mixed with 1 ml of $CoCl_2$ (3 mg mL^{-1}) using sonication. It should be noted, for synthesis a very diluted amount of metal salt or organometallic compound always used. Resultant materials could have bigger metal nanoparticles along with a single-atom catalyst, if excess amount is used. Hence, atomic dispersion and number of metal active sites can be controlled using the optimized amount of metal precursor. Then, the whole mixer was freezing dried and heated at 750°C for 1 h underflow of NH_3 gas (50 cm^3/min) and argon (150 mL/min). Similarly, a control catalyst was prepared for comparison purposes. HAADF, and EXAFS surface characterization and electrochemical measurements have suggested the formation of Co-N_4 sites (**Figure 5**).

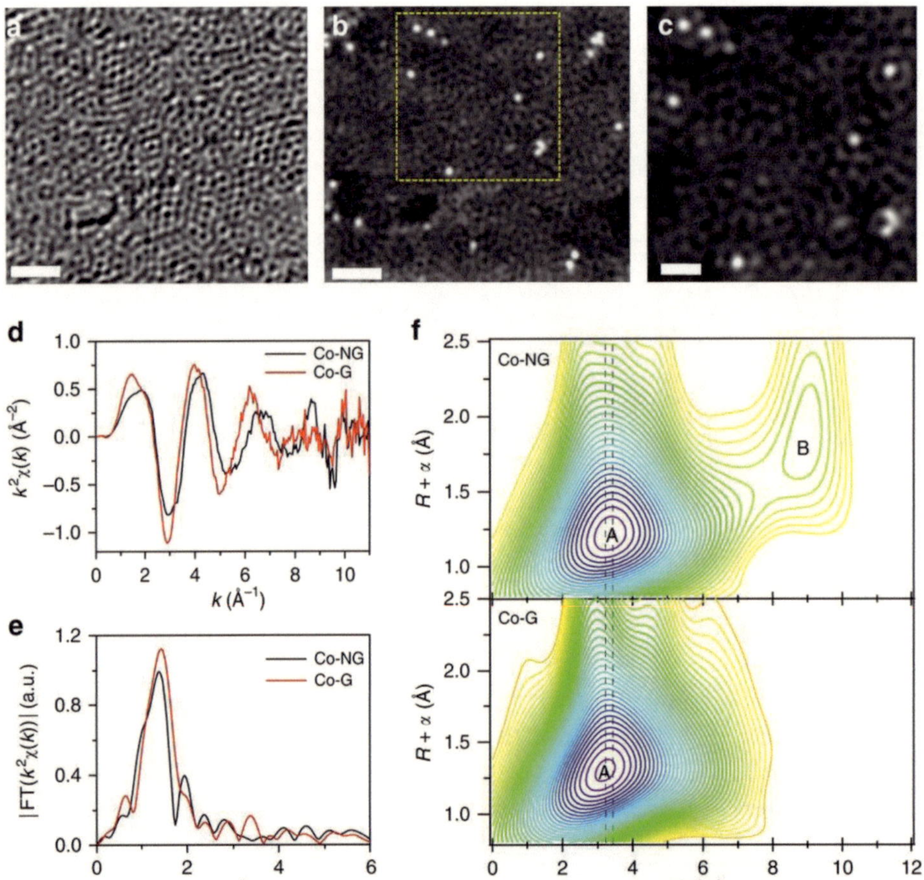

Figure 5.
Structural characterization of Co-NG catalyst. (a) Bright filed aberration-corrected STEM images, scale bar 1 nm. (b) HAADF-STEM image. (c) Enlarge view of a selected area in image b, scale bar 0.5 nm. (d, e) K_2 weighted EXAFS in K space and their Fourier transformation in T space for the Co-NG and Co-G, respectively. (f) Wavelength transformation of Co-NG and Co-G (Reproduced with the permission, 2015, Nature Publishing Group [19]).

All these data indicating that Co is atomically dispersed in nitrogen-doped graphene support (**Figure 5**). Bond distance calculation clearly revealed the formation of a Co-N bond (**Figure 5d–f**). This proved that *N*-atom in graphene support provide anchoring to Co atom and did not allow to agglomerate during catalytic cycles. After that, Co-NG was checked for HER catalytic activity using standard three-electrode electrochemical cell. A glassy electrode was prepared and loaded with 285 μg cm^{-2} catalyst. **Figure 4a** showed LSVs curves of the samples which were recorded at 2 mV s^{-1} scan rate in 0.5 M H_2SO_4 aqueous solution. Co-NG catalyst showed excellent HER catalytic activity with a 30 mV onset potential at a current density of −0.3 mA cm^{-2}, where in comparison commercial available Pt catalyst showed zero onset potential as expected. Onset potential was 70 and 147 mV at 1 and 10 mA cm^{-2} current density, respectively. **Figure 4b** showed 100% Faradaic efficiency, which was due to the generation of H_2. Faradaic efficiency calculated based on amount of charge transferred through an electrochemical system and amount of reaction happened. In their work, authors were able to achieve better onset potential compared with the Co-based molecular complexes. Further comparison with control catalysts nitrogen-doped graphene and Co-G, single-atom Co-NG catalyst performed much better (**Figure 4c**), which clearly indicated the role of *N* atom graphene matrix. These catalysts were also tested in alkaline media (1 M NaOH) solution. Single-atom Co-NG catalyst also showed improved catalytic activity compared with the Co-G, NG, and some of other catalysts such as MoS_2 and Ni_2P, were found unstable in a base solution (**Figure 4d**).

TOF is the most important parameter to compare the catalytic activity per unit active sites. In Co-NG, every single atom considered as the active site and if contribution from the NG matric ignored, the exchange current density by Tafel plots, Co-NG showed 1.25×10^{-4} A cm^{-2} which is much better than NG (8.43×10^{-7} A cm^{-2}). To evaluate the stability of the catalyst, accelerated degradation experiments in both acidic and basic media were carried out. One thousand continuous cathodic polarization CV curves showed the excellent stability of the Co-NG compared with the other controlled samples. In addition to the 1000 CV cycles, galvanostatic measurements at a current density of 10 mA cm^{-2} were showed the 35 and 15 mV decrease in offset value after 10 h continuous operation.

2.2 Oxygen reduction reactions (ORR)

ORR is a common relevant process in many electrochemical energy conversion and storage technologies including the air batteries and fuel cells. In most of the cases, ORR reaction is carried out in alkaline conditions, because ORR is much more facile and also provides the high activity in non-noble metal catalyst including TM oxide, SMACs, and heteroatoms doped nanostructured carbon. ORR reaction also proceeds with two electrons reduction mechanism in aqueous medium.

Acidic medium:
Direct pathway: $O_2 + 4H^+ + 4e^- \rightarrow 2H_2O \quad E_0 = 1.23$ V
Series pathways: $O_2 + 4H^+ + 4e^- \rightarrow H_2O_2 \quad E_0 = 0.69$ V
$H_2O_2 + 2H^+ + 2e^- \rightarrow 2H_2O \quad E_0 = 1.23$ V

Alkaline medium:
Direct pathway: $O_2 + 2H_2O + 4e^- \rightarrow 4OH^- \quad E_0 = 1.23$ V
Series pathways: $O_2 + H_2O + 2e^- \rightarrow H_2O^- + OH^- \quad E_0 = 0.69$ V
$HO_2^- + H_2O + 2e^- \rightarrow 3OH^- \quad E_0 = 1.23$ V

Yang and co-workers produced a single Co atom using N doped carbon nanofiber for ORR application [24]; 1 g of polyacrylonitrile powder was dissolved in 12.5 g of dimethylformamide under stirring at 60°C for 2 h. Then, 2 g of 4-dimethylamino pyridine powder and 150 mg of cobalt acetate were added and stirred further at room temperature for 2 h. The paste was used to electrospinning at 35°C. The film was dried under vacuum at 60°C for 12 h and then stabilized by annealing at 250°C for 2 h in the air. Finally, Co-containing film was heat-treated at 900°C for 2 h under inert gas condition. As prepared, sample could have bigger Co nanoparticles along with the single atom embedded in the carbon framework. Samples were treated with the aqueous solution of 5 M H_2SO_4, washed and dried under vacuum at 60°C. The authors also prepared nitrogen-doped carbon fiber and only carbon fiber for comparison.

TEM imaging (**Figure 6A–F**), EDX mapping revealed the complete dispersion of Cobalt in carbon fiber. HAADF-STEM images confirmed the formation of a single Co atom and there was no signature of bigger Co nanoparticles formation observed (**Figure 6A–F**). XANES and EXAFS were used to investigate the local structure of a single Co atom in the *N* doped carbon nanofiber. XANES and EXAFS spectroscopy revealed the formation Co-N bond in a carbon matrix, also conforming to the existence of a single Co atom [24].

The catalyst was used to find out its ORR catalytic activity in 0.1 M $HClO_4$ solution and showed an onset potential of 0.82 and 0.70 V versus the reversal hydrogen electrode (RHE) (**Figure 6a**). Author found as prepared catalyst showed good catalytic activity, however it was still lower than Pt/C with an onset potential of 0.94 and 0.83 V, respectively. In comparison with the other control catalyst, the catalytic order was CNF < NCNF < Co-N-CNF, which confirmed the important role of *N* in the carbon matrix to anchor Co single atom. The catalyst was found stable in both acidic and basic media, showed high durability (**Figure 6c**), high catalytic activity and stability (**Figure 6f**). The half-wave potential of ORR reaction did not show many changes even after the 10,000 cycles. This is might be due to structural stability of atomically dispersed Co atom in the *N* doped carbon network [24].

To further clarify unique effect of single Co atom on ORR activity, authors prepared Co nanoparticles and *N* doped CoP-N/CNFs. It clearly showed 34 mV negative shifts in half-wave potential after 1000 cycles. This is attributed to the stability of single Co atoms anchored on *N* atom in carbon framework. Now next question is what is chemical nature of Co atom attached to *N* and expected it can be Co-N_4 as it was proved multiple times in the case of Fe single-atom catalyst. DFT theory was used to prove the nature of active sites on the basis of previously reported (transition metal) TM-N_4-graphene (G) model. **Figure 7a** showed the ion exchange energy landscapes for the Co-N_4-G as well for Fe-N_4-G after the adsorption of molecular and atomic oxygen under acidic conditions. After geometry optimization, O_2 absorption energy in parallel and end on a model for Fe-N_4-G and Co-N_4-G respectively (**Figure 7a**). Structure stability of Co-N_4/CNF can be explained on basis of free energy diagrams of the ORR as showed in **Figure 7b**. The structural stability of Co-N_4/CNF is further increasing after O_2 and O adsorption, which could be the reason for increase in structure stability of the catalyst. Catalyst also showed much better performance for direct methanol fuel cells with a maximum power density of 16 mW cm^{-2} for 200 h (**Figure 7c**). The lifetime test of the passive DMFC showed a nearly constant cell voltage of 30 V after the discharge of 200 h (**Figure 7d**), indicating excellent stability of the electrode. This could be used for a robust fuel cell application in the future [24].

Li and co-workers reported the synthesis of a single Co atom catalyst with precise *N* coordination as superior ORR catalyst [25]. For synthesis, author used Zn/Co bimetallic MOFs, during the pyrolysis process, Zn can selectively be evaporated away and remaining have a carbon framework decorated with a single Co atom. In a typical procedure, $Co(NO_3)_2$ and $Zn(NO_3)_2$ was dissolved in 15 mL methanol and which was later mixed with 2-methylimidazole in methanol at room temperature.

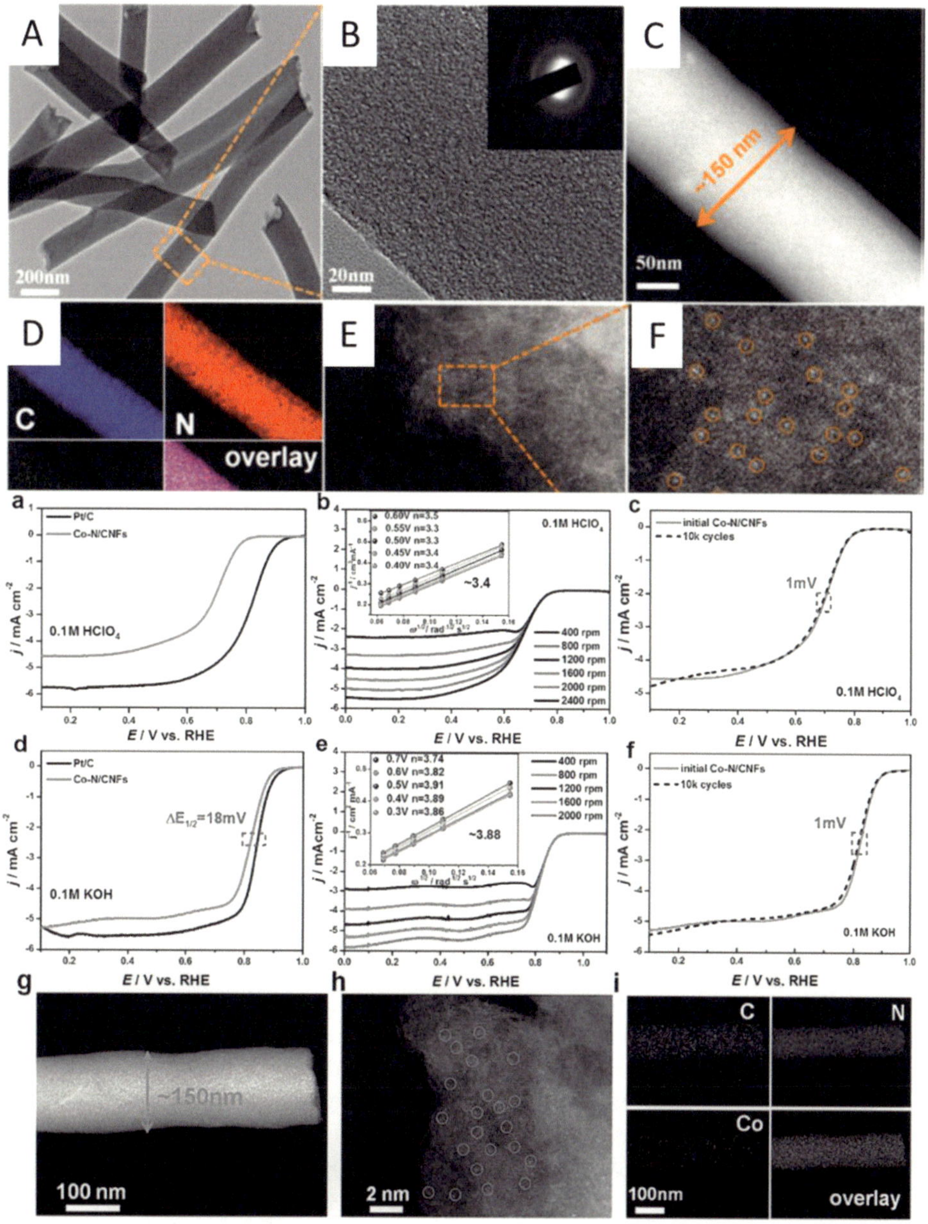

Figure 6.
(A) TEM images of fibers, (B) high magnification TEM images of Co-N/CNF, (C) HAADF-STEM images, (D) EDX mapping, (E) aberration-corrected STEM images, (F) enlarge image of Co-N/CNF (single Co atom were highlight by orange circle). (a) LSV of ORR on the Co-N/CNF and Pt/C. (b) K-L plots derived from LSV of ORR. (c) Durability test. (d) LSV of ORR on the Co-N/CNF and Pt/C in basic condition. (e) K-L plots derived from LSV of ORR. (f) Durability test. (g) STEM images of single nanofiber. (h) Aberration-corrected STEM images. (i) EDX mapping for Co-N/CNF after a 1000 cycles ADT test (Reproduced with the permission, 2017, American Chemical Society [24]).

The mixed solution was then transferred to 50 mL hydrothermal lined autoclave and heated at 120°C for 4 h. As prepared precipitates were then centrifuged and washed with ethanol several times and dried in vacuum at 70°C for overnight. For carbonization, as prepared MOF was then heated at 800, 900, and 1000°C for 3 h under the flow of N_2 gas. Then heated MOFs were then used directly for electrochemical measurement without further treatment.

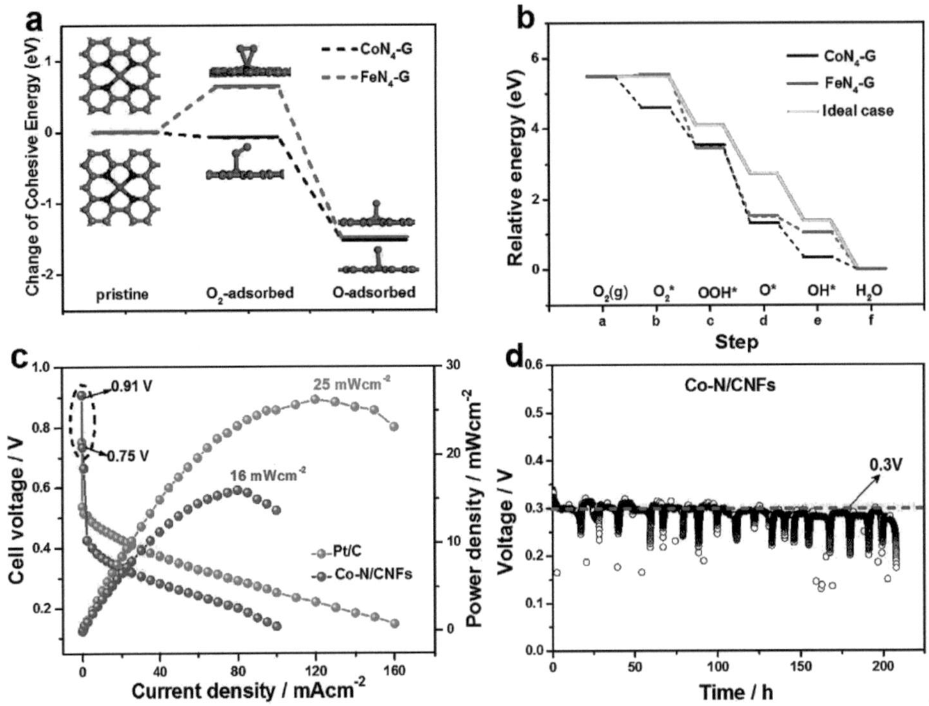

Figure 7.
(a) Relative ion-exchange energy profile. (b) Relative energy profile of the ORR process of the active sites. (c) Steady-state polarization curve of passive DMFCs with 4 M methanol at 40°C. (d) 200 h stability test of a passive DMFC with the cathode catalyst prepared by Co-N/CNFs at a current density of 20 mA cm^{-2} (Reproduced with the permission, 2017, American Chemical Society [24]).

MOFs were used as a starting precursor for the synthesis because MOF already has distinguished single metal sites coordinated with the organic ligand. Mixed metallic MOF used because some portion of Co is in pure Co MOF can be replaced with the low boiling point Zn atom, which further helps in maintaining the appropriate distance between the cobalt atoms in the final carbonized materials. Co loading can be tuned by the changing ratio of metal salt added during MOF synthesis. HAADF-STEM images revealed the formation of homogeneous Co atom on carbon support [25].

The special aberration correction spectroscopy and extended X-ray absorption fine structure techniques have been used to confirm the atomic dispersion of Co atom anchored by *N* atom in the carbon framework [25]. ORR activity of the catalyst was studied using steady-state linear sweep voltammetry (LSV) on the rotating disk electrode in O_2 saturated 0.1 M KOH solution. For comparison, authors also prepared *N* doped catalyst by pyrolysis of Zn MOFs. This showed low ORR activity, exhibits the role of a single Co atom for its excellent ORR activity. Co-SAs/N-C(900) showed the best ORR activity in comparison with another catalysts, onset potential of 0.881 V was almost equal to the Pt/C catalyst (0.811). This was found in correlation with the DFT calculation. KL plots of Co-SAs/N-C(900) suggesting first-order reaction kinetic. Electron transfer number has been calculated using the rotating ring-disk electrode (RRDE) tests, showing the nearly four-electron ORR pathways over the single Co-N_2 sites. The kinetic current density (J_k) by the Tofel plots at 0.8 V was around 75 mV dec^{-1}. The stability test showed that the catalyst is stable during the operational condition, no change noticed in the CV curve even after 5000 cycles.

Abruna *et al*. produced structurally disordered intermetallic platinum cobalt core-shell nanoparticles for ORR activity [26]. They demonstrated Pt-metal disordered alloy core-shell nanoparticles with ordered Pt_3Co intermetallic core and 2–3 layer of Pt shell. These Pt-Co core-shell nanoparticles showed a 200% increase in mass activity and 300% increase in specific activity when compared with disordered Pt_3Co alloy nanoparticles as well as with Pt/C. The stability test showed minimal loose of ORR activity after 5000 cycles and core-shell structure was still found stable under operational conditions. Lin *et al*. reported the preparation of a hierarchically porous Co and Fe single-atom catalyst for ORR [27]. In present work, authors used bimodal template bases synthesis strategy for the large scale synthesis of porous single-atom catalysts. For synthesis of hierarchically porous Co-N-C-900 single-atom catalyst, first 2 g D-glucosamine hydrochlorides, $ZnCl_2$, and $CoCl_2$ were dissolved in 20 mL colloidal silica suspension. After stirring for 10 min, whole mixture was then freeze-dried and obtained powder was carbonized under the flow of N_2 gas at 900°C for 2 h. Silica was removed using 10 wt.% HF solution for 12 h at room temperature. For comparison, Co-N-C-900 was synthesized without using $ZnCl_2$ and also porous N-C was also synthesized. Silica was used as a hard template because this creates the porosity in a carbon network and further use of $ZnCl_2$ produced more nanopores in carbon networks. TEM analysis clearly showed the presence of nanopores of 20 nm size, due to the use of silica nanoparticles as a template. HAADF-STEM images of catalyst represent the formation of a single atom in a carbon framework. No bigger nanoparticles were detected, this is because during HF treatment, it could be possible for bigger particles to dissolve and only those single atoms remain which are strongly bound with the carbon framework. Fe single-atom catalyst was found stable against O_2 and also found most active compared with Co single-atom catalyst. Optimized Fe-N-C catalyst showed half-wave potential of 0.927 V in alkaline medium, which was found even more positive 49 and 55 V than the C0-N-C and Pt/C, respectively. DFT calculation revealed that Fe-N_2 sites are more active compared with the CoN_4 for ORR because of Fe-N_2 functional at lower energy barrier for the intermediate and produced involved.

Sun *et al*. reported synthesis of a single Co atom embedded in hierarchically ordered porous nitrogen-doped carbon (Co-SAS/HOPNC) for HER and ORR applications [28]. Firstly, phenol-formaldehyde resol (Mw $<$ 500) was prepared using 10 g phenol and formaldehyde. For the synthesis of Co-SAS/HOPNC, pluronic F127 and dicyandiamine were dispersed in ethanol and deionized water using magnetic stirrer, followed by adding a resol ethanol solution. Then vitamin B12 protein was added to above mixture at room temperature to produce a homogeneous pink solution. Subsequently, silica template was soaked into the prepared precursor for 2 h and whole mixture heated at 50°C in an electric oven for solvent evaporation for 8 h. After heating at 100°C for overnight for thermosetting, as prepared composite was heated at 900°C for 2 h in an Ar atmosphere. To remove the bigger cluster of Co nanoparticles and silica after carbonization; composite was treated with the HF aqueous solution for 24 h. As prepared catalyst possessed high surface area, good conductivity, high isolated Co atom, and exceptional catalytic performance. Co-SAS/HOPNC exhibited the half-wave potential of 0.892 V for ORR in alkaline medium. This was found 53 mV more positive than Pt/C, high tolerance to methanol and great stability. Co-SAS/HOPNC showed remarkable HER activity with a turnover frequency of 0.41 and 3.8 s^{-1} at overpotential of 100 and 200 mV, respectively. DFT theory was proposed to explain the nature of single Co site. Co-N_4 was found an active site for this dual character, three-dimensional porous carbon networks; accessible highly active Co-N_4 sites synergistically contribute the excellent catalytic performance.

3. Conclusions and future directions

Single-atom catalyst has established a connection between homogeneous and heterogeneous processes. A single atom catalyst can be better than a homogeneous catalyst because it is easy to recover at the end of the process and equal catalytic activity. Single-atom catalyst can have many fascinating characteristics such as maximize the atomic efficiency, a high activity like a homogeneous system, and selectivity. A single-atom catalyst has been proved to be a very efficient and powerful catalyst for various applications including the ORR, OER, HER, CO_2 conversion, and organic transformation. Significant progress is made to understand the catalyst preparation, characterization using advanced techniques and mechanism interpretation using experimental data and theoretical models.

The microenvironment of a single atomic site needs to investigate more in detail, *in situ* spectroscopy technique can be utilized to study the active site in operation condition, and a specific reaction mechanism can be proposed [29]. It can be understood from literature, the interaction between the metal single atom and support is necessary to stabilize a single metal atom. If we can manipulate these interactions, it can tune the catalytic activity, stability, and selectivity for multiple catalytic applications.

MOFs are the fascinating precursor for single metal atom catalyst design. Multimetal MOFs can also use for the synthesis of multi-metallic single atom heterogeneous catalyst to utilized the one catalyst for numerous applications [30]. To further change the catalytic phenomena of a single metal catalyst, two-dimensional materials such as MXenes, MoS_2, HBN, black phosphorus, and many more support materials can be successfully explored for the synthesis, industrial, and environmental applications [31].

Porous metal oxide support for the preparation of a single-metal atom catalyst can be also utilized and use for organic transformation, gas phase, and electrochemical applications. Although porosity of metal oxide support is an issue, however two-dimensional metal oxides can be utilized for single-atom catalyst preparation [32]. Silica (mesoporous spherical or nanosheets) can be a good support for the single-atom catalytic application, however, it needs to be explored further [33, 34]. Non-conducting and passive nature of silica might a problem for single-atom catalysts. Porous carbon nanocage using KCC-1 as a hard template was used to prepare a three-dimensional accessible surface carbon sphere [35]. Same carbon morphology can be utilized for the preparation of mesoporous carbon-based SAMCs for various applications. **Figure 8** is demonstrating the overall picture of different ways of single-atom catalyst preparation and their utilization in various applications.

Recently, metal-air batteries have been received to consider a potential candidate for future energy storage materials due to their theoretical high energy density, excellent safety, and abundant TM resources. Typically, metal-air batteries' performance

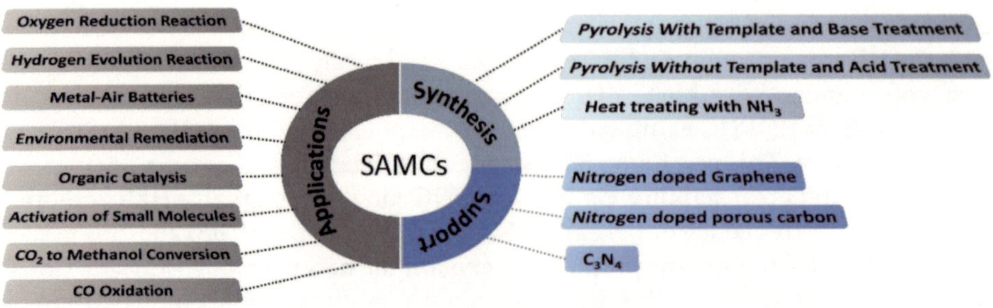

Figure 8.
General discerption of SMACs, method of preparation, support used and applications.

depends on two catalytic reactions ORR and OER, both occurring on porous cathode surface [36–38]. Due to the unsaturated characteristic of SMACs and its catalytic activity can be utilized to make more highly efficient and cheap metal-air batteries.

Acknowledgements

We thank SERB - DST, New Delhi for their support under EMEQ Scheme.

Conflict of interest

Authors declare that there is no conflict of interest in this chapter.

Acronyms and abbreviations

ORR	oxygen reduction reaction
OER	oxygen evolution reaction
HER	hydrogen evolution reaction
CUMS	coordinately unsaturated metal sites
SMACs	single metal atom catalysts
MM-SACs	multi-metallic single-atom catalyst

Author details

Baljeet Singh[1*], Surender Kumar[2*] and Archana Singh[2]

1 Tata Institute of Fundamental Research (TIFRH), Hyderabad, India

2 CSIR - Advanced Materials and Processes Research Institute, Bhopal, India

*Address all correspondence to: itsbaljeet5@gmail.com
and surenderjanagal@gmail.com

References

[1] Yang XF, Wang AQ, Qiao BT, Li J, Liu JY, Zhang T. Single-atom catalysts: A new frontier in heterogeneous catalysis. Accounts of Chemical Research. 2013;**46**(8):1740-1748

[2] Wang Y, Mao J, Meng XG, Yu L, Deng DH, Bao XH. Catalysis with two-dimensional materials confining single atoms: Concept, design, and applications. Chemical Reviews. 2019;**119**(3):1806-1854

[3] Su X, Yang XF, Huang YQ, Liu B, Zhang T. Single-atom catalysis toward efficient CO_2 conversion to CO and formate products. Accounts of Chemical Research. 2019;**52**(3):656-664

[4] Wang CS, Wang H, Wu RQ, Ragan R. Evaluating the stability of single-atom catalysts with high chemical activity. Journal of Physical Chemistry C. 2018;**122**(38):21919-21926

[5] Zhang XF, Guo JJ, Guan PF, Liu CJ, Huang H, Xue FH, et al. Catalytically active single-atom niobium in graphitic layers. Nature Communications. 2013;**4**

[6] Wang AQ, Li J, Zhang T. Heterogeneous single-atom catalysis. Nature Reviews Chemistry. 2018;**2**(6):65-81

[7] Liu WP, Sahoo B, Junge K, Beller M. Cobalt complexes as an emerging class of catalysts for homogeneous hydrogenations. Accounts of Chemical Research. 2018;**51**(8):1858-1869

[8] Sahoo S, Suib SL, Alpay SP. Graphene supported single atom transition metal catalysts for methane activation. ChemCatChem. 2018;**10**(15):3229-3235

[9] Xue YR, Huang BL, Yi YP, Guo Y, Zuo ZC, Li YJ, et al. Anchoring zero valence single atoms of nickel and iron on graphdiyne for hydrogen evolution. Nature Communications. 2018;**9**

[10] Yan H, Su CL, He J, Chen W. Single-atom catalysts and their applications in organic chemistry. Journal of Materials Chemistry A. 2018;**6**(19):8793-8814

[11] Cui XJ, Li W, Ryabchuk P, Junge K, Beller M. Bridging homogeneous and heterogeneous catalysis by heterogeneous single-metal-site catalysts. Nature Catalysis. 2018;**1**(6):385-397

[12] Zhao J, Deng QM, Avdoshenko SM, Fu L, Eckert J, Ruemmeli MH. Direct in situ observations of single Fe atom catalytic processes and anomalous diffusion at graphene edges. Proceedings of the National Academy of Sciences of the United States of America. 2014;**111**(44):15641-15646

[13] Markevich AV, Baldoni M, Warner JH, Kirkland AI, Besley E. Dynamic behavior of single Fe atoms embedded in graphene. Journal of Physical Chemistry C. 2016;**120**(38):21998-22003

[14] Zhao L, Zhang Y, Huang LB, Liu XZ, Zhang QH, He C, et al. Cascade anchoring strategy for general mass production of high-loading single-atomic metal-nitrogen catalysts. Nature Communications. 2019;**10**

[15] Liu JY. Catalysis by supported single metal atoms. ACS Catalysis. 2017;**7**(1):34-59

[16] Zhang H, Liu G, Shi L, Ye J. Single-atom catalysts: Emerging multifunctional materials in heterogeneous catalysis. Advanced Energy Materials. 2018;**8**(1):1701343

[17] Zhao GX, Pang H, Liu GG, Li P, Liu HM, Zhang HB, et al. Co-porphyrin/

carbon nitride hybrids for improved photocatalytic CO_2 reduction under visible light. Applied Catalysis B: Environmental. 2017;**200**:141-149

[18] Li X, Bi W, Zhang L, Tao S, Chu W, Zhang Q, et al. Single-atom Pt as Co-catalyst for enhanced photocatalytic H_2 evolution. Advanced Materials. 2016;**28**(12):2427-2431

[19] Fei HL, Dong JC, Arellano-Jimenez MJ, Ye GL, Kim ND, Samuel ELG, et al. Atomic cobalt on nitrogen-doped graphene for hydrogen generation. Nature Communications. 2015;**6**

[20] Peng Y, Lu B, Chen S. Carbon-supported single atom catalysts for electrochemical energy conversion and storage. Advanced Materials. 2018;**30**(48):1801995

[21] Aiyappa HB, Masa J, Andronescu C, Muhler M, Fischer RA, Schuhmann W. MOFs for electrocatalysis: From serendipity to design strategies. Small Methods. 2019;**3**(8)

[22] Chen ZP, Pronkin S, Fellinger TP, Kailasam K, Vile G, Albani D, et al. Merging single-atom-dispersed silver and carbon nitride to a joint electronic system via copolymerization with silver tricyanomethanide. ACS Nano. 2016;**10**(3):3166-3175

[23] Liang HW, Wei W, Wu ZS, Feng XL, Mullen K. Mesoporous metal-nitrogen-doped carbon electrocatalysts for highly efficient oxygen reduction reaction. Journal of the American Chemical Society. 2013;**135**(43):16002-16005

[24] Cheng QQ, Yang LJ, Zou LL, Zou ZQ, Chen C, Hu Z, et al. Single cobalt atom and N codoped carbon nanofibers as highly durable electrocatalyst for oxygen reduction reaction. ACS Catalysis. 2017;**7**(10):6864-6871

[25] Yin P, Yao T, Wu Y, Zheng L, Lin Y, Liu W, et al. Single cobalt atoms with precise N-coordination as superior oxygen reduction reaction catalysts. Angewandte Chemie, International Edition. 2016;**55**(36):10800-10805

[26] Wang DL, Xin HLL, Hovden R, Wang HS, Yu YC, Muller DA, et al. Structurally ordered intermetallic platinum-cobalt core-shell nanoparticles with enhanced activity and stability as oxygen reduction electrocatalysts. Nature Materials. 2013;**12**(1):81-87

[27] Zhu C, Shi Q, Xu BZ, Fu S, Wan G, Yang C, et al. Hierarchically porous M–N–C (M = Co and Fe) single-atom electrocatalysts with robust MNx active moieties enable enhanced ORR performance. Advanced Energy Materials. 2018;**8**(29):1801956

[28] Sun TT, Zhao S, Chen WX, Zhai D, Dong JC, Wang Y, et al. Single-atomic cobalt sites embedded in hierarchically ordered porous nitrogen-doped carbon as a superior bifunctional electrocatalyst. Proceedings of the National Academy of Sciences of the United States of America. 2018;**115**(50):12692-12697

[29] Wang J, Gan LY, Zhang WY, Peng YC, Yu H, Yan QY, et al. In situ formation of molecular Ni-Fe active sites on heteroatom-doped graphene as a heterogeneous electrocatalyst toward oxygen evolution. Science Advances. 2018;**4**(3)

[30] Zheng FQ, Xiang D, Li P, Zhang ZW, Du C, Zhuang ZH, et al. Highly conductive bimetallic Ni-Fe metal organic framework as a novel electrocatalyst for water oxidation. ACS Sustainable Chemistry & Engineering. 2019;**7**(11):9743-9749

[31] Li Z, Zhang X, Cheng H, Liu J, Shao M, Wei M, et al. Confined

synthesis of 2D nanostructured materials toward electrocatalysis. Advanced Energy Materials:1900486

[32] Liu JC, Wang YG, Li J. Toward rational design of oxide-supported single-atom catalysts: Atomic dispersion of gold on ceria. Journal of the American Chemical Society. 2017;**139**(17):6190-6199

[33] Bayal N, Singh B, Singh R, Polshettiwar V. Size and fiber density controlled synthesis of fibrous nanosilica spheres (KCC-1). Scientific Reports. 2016;**6**

[34] Singh B, Polshettiwar V. Solution-phase synthesis of two-dimensional silica nanosheets using soft templates and their applications in CO_2 capture. Nanoscale. 2019;**11**(12):5365-5376

[35] Singh B, Maity A, Polshettiwar V. Synthesis of high surface area carbon Nanospheres with wrinkled cages and their CO_2 capture studies. ChemistrySelect. 2018;**3**(38):10684-10688

[36] Brij K, Baljeet S, Surender K. Graphene-based nanocatalysts for oxygen reduction and evolution reactions in metal-oxygen batteries. Current Catalysis. 2018;**7**(3):158-166

[37] Liu MM, Wang LL, Zhao KN, Shi SS, Shao QS, Zhang L, et al. Atomically dispersed metal catalysts for the oxygen reduction reaction: Synthesis, characterization, reaction mechanisms and electrochemical energy applications. Energy & Environmental Science. 2019;**12**(10):2890-2923

[38] Peng P, Shi L, Huo F, Mi CX, Wu XH, Zhang SJ, et al. A pyrolysis-free path toward superiorly catalytic nitrogen-coordinated single atom. Science Advances. 2019;**5**(8)

Section 6

Perovskite-Type Material Lanthanum Cobaltite $LaCoO_3$: Aspects of Processing Route Toward Practical Applications

Chapter 6

Perovskite-Type Lanthanum Cobaltite $LaCoO_3$: Aspects of Processing Route toward Practical Applications

Mirela Dragan, Stanica Enache, Mihai Varlam and Konstantin Petrov

Abstract

Lanthanum cobaltite ($LaCoO_3$) perovskite-type oxide is an important conductive ceramic material finding a broad range of technical applications. Physical and chemical properties of the final lanthanum cobalt oxide powder material obtained are strongly dependent on the method of preparation. Taking in account these considerations, we focus our investigation on the solid state reaction process. The characterization of prepared lanthanum cobalt oxide material was studied by using X-ray diffractometry (XRD), scanning electron microscopy (SEM), thermogravimetry-differential scanning calorimetry (TG-DSC), and conduction properties. Following the experimental results, it can be concluded that with proper improvement, the solid state reaction process may also provide an efficient preparation method for perovskite-type $LaCoO_3$ powder. Important to mention is that we looked into the aspects to produce again same which showed consistently reproducibility of batch to batch powder properties. This is a key factor to overcome a successful commercialization of new material synthesis development.

Keywords: perovskite, $LaCoO_3$, oxide powders synthesis, solid state reaction

1. Introduction

The continuous interaction between structure and properties allows several intrinsic properties of perovskite materials to advance a very broad range of practical applications. These oxides are being increasingly applied to electronic and magnetic materials [1–3], automotive exhaust and water splitting catalysts [4, 5], and electrode materials for fuel cells and batteries [6, 7]. Among these perovskites, the cobalt-based type $LaCoO_3$ perovskite ignited interest in the research since the early 1960s [8, 9] and continues to be the material of the moment. $LaCoO_3$ perovskite has been shown to have promising catalytic activity for oxygen evolution reaction (OER) [10, 11]. Lanthanum, La, is a relatively large cation and gives structural stability to the catalyst. Rare-earth oxides with full or partially filled inner shells of lanthanide ions would involve the 4f electrons. The 4f electrons contribute to the density of states around the Fermi level, and degenerate strongly, the bandwidth is narrow and steep. The presence of La^{3+} with no 4f electrons is beneficial because the electrical conductivity increases and the effective mass decreases [12].

Cobalt, Co, cation, a transition metal with smaller size than La cation, is responsible for catalytic activity. In its compounds, cobalt nearly always exhibits a +2 or +3 oxidation state, although states +4, +1, 0, and −1 are also known [13]. The outer electrons of the element are either in the 3d or 4s subshell. Cobalt chemistry is dominated by the behavior of the 3d electrons. Oxidation state of Co helps the $LaCoO_3$ catalytic activity for OER which can be associated with Co^{3+} oxidation state. Co^{3+} active sites by absorption of HO^- may act as reactants for OER [14]. Electrocatalysis at room temperature, in correlation to the fundamental electronic structure, is still not fully clarified. For Co_3O_4 with the band gap of 1.9 eV and $LaCoO_3$ with about 0.8 eV, the resistivity is 10^4 and 10 Ωcm, respectively [15, 16].

On the other hand, crystallographic structure is important for the functional properties of the oxides. For perovskite, the ideal cubic structure goes through different structural distortions due to the ionic radii differences.

The equation determined by Goldschmidt correlates crystal structures geometrically in terms of the ionic packing using the Goldschmidt's tolerance factor t. For a stable perovskite structure, the tolerance factor should have values between 0.75 and 1; otherwise, the ideal cubic structure is a distorted structure. Mathematical expression involving the unit cell length ratio, here r_{La}, r_{Co}, and r_O, is the ionic radii for La, Co, and O respectively; t is given as:

$$t = \frac{(r_{La} + r_O)}{\sqrt{2}\,(r_{Co} + r_O)} \tag{1}$$

Taking in consideration the Shannon crystal ionic radii for all ions along with their coordination numbers [17], the calculated tolerance factor of $LaCoO_3$ is 0.97 for Co^{3+} and 0.905 for Co^{2+}.

Experimentally, by neutron diffraction technique, the change of lattice parameter with temperature has been observed when the lattice length and angle become longer and smaller, respectively, when temperature increases [18].

The crystal structure of $LaCoO_3$, as shown in **Figure 1**—La atoms, in blue, at corners; O atoms, in red, on face centers; and Co atoms, in green, at the center of the lattice—is a rhombohedron having R3c symmetry at room temperature. This is considered as the most stable: $LaCoO_3$ structure up to around 1698 K when the crystal is cubic [19].

Not only the temperature but also the oxygen partial pressure of environments, $p(O_2)$ leads the $LaCoO_3$ to a wide range of oxygen deficiency represented by the formula $LaCoO_{3-\delta}$ with δ being the oxygen deficiency. It is observed that moderate oxygen deficiency in $LaCoO_{3-\delta}$ causes a slight distortion of the ideal cubic structure which is rombohedral [20].

Since the structure-properties relations are strongly dependent on the preparation method, a lot of interest is focused on their synthesis which ultimately determines its potential applications.

The aim of this research is to explore the possibility to improve the solid state preparation method for obtaining $LaCoO_3$ micro- and nanocrystals, which may contribute to the development of a large-scale production route of $LaCoO_3$ with controlled properties.

Numerous routes to prepare perovskite powders have been proposed. These address various problems involved in the preparation of polycrystalline perovskite powder with single phase, resulting in various microstructures and properties.

Among the adopted techniques, solid state reaction [21], mechanochemical processing [22], Pechini method [23], combustion synthesis [24], sol-gel method [25], and microwave route [26] are widely used.

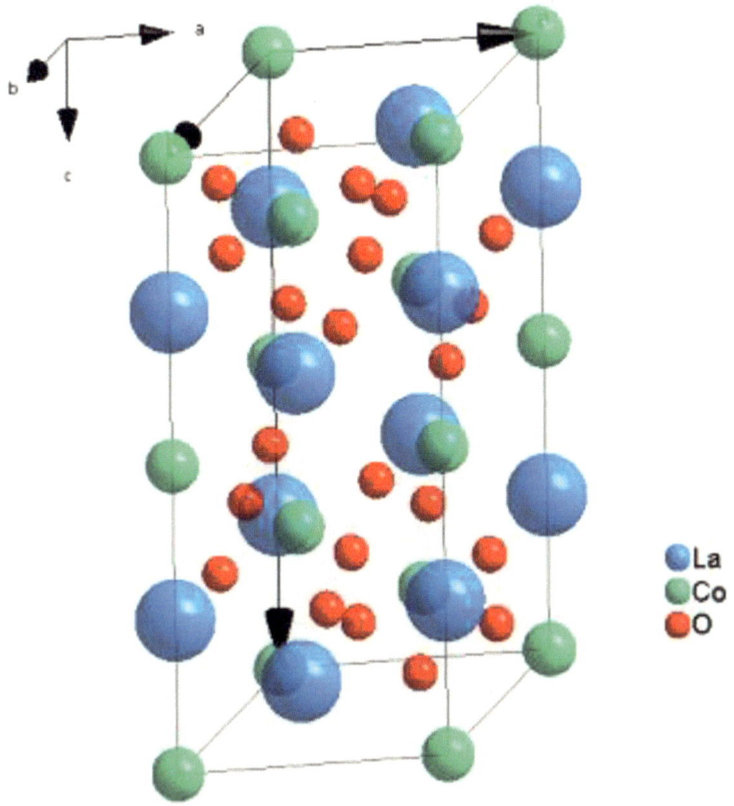

Figure 1.
The crystal structure of rhombohedral $LaCoO_3$.

Despite these various methods, some technological applications are still limited for $LaCoO_3$ compounds. Therefore, knowledge of the preparation conditions and optimization of the technical parameters are key challenges.

The advantages like low temperatures used during processing of wet preparation routes lead to oxide perovskite powders with high-purity nanoparticles. However, these techniques cannot achieve full industrial potential. For instance, major drawbacks of these methods are the high costs of raw materials, the large quantity of gaseous by-products and waste products such as nitrates, acetates, and additional chelating agents used as precursors. The use of such materials requires elaborate systems for gas collection and storage since the by-products are poisonous. Their significant toxicity is not environmentally friendly, making it not sustainable for large-scale applications. In addition to that, these methods involve many stages during the synthesis process. One way to circumvent that is to seek solutions for adapting clean processing routes to specific needs of the products without altering the physical and chemical properties of the resulting powders. For instance, if particle-size distribution is not especially required for certain applications, solid-state synthesis route has superior advantage over wet processing routes owing to its simplicity, less equipment requirements, and ease of scaling up for industrial production.

The elimination of wet processing for perovskite powder preparation means that products will cost less both ways, to produce as well to buy, and have an important economic advantage. In addition, by mixing dry powdered precursors in the stoichiometric ratios followed by heating to obtain the desired reaction product, at the end of the reaction, there is no waste to dispose. In this way, the method is also environmentally friendly.

The overall price of oxide perovskite powders is strongly influenced by the industrial production route. Solid state reaction technology is resource- and energy-saving. The processing route allows one to obtain the reaction products readily in one step by using less expensive precursors, without waste products. The upscale of solid-state synthesis route resumes to batch production. This is different from flow processing which is a continuous production technique that has inflexibility to adjust if the case. Additionally, the absence of waste and by-products is another major positive aspect regarding the environmental impact of the entire production process. This is why we consider this method useful for both environmental protection and industrial use.

In solid state reaction, rates are typically diffusion-limited; consequently, decreasing the diffusion distances through intimate mixing of reactants possibly contributes to overcome the diffusion barrier in addition to high temperature during the thermal treatment. The characteristics of the starting materials impact the area of contact between the reacting particles with influence of the rate of reaction.

During the heating regime, a non-isothermal transformation takes place, and during the dwell regime, an isothermal transformation takes place. A significant kinetic parameter for studying the phase transformation of precursors is the activation energy E_a, representing the energy barrier for atoms and molecules to move and rearrange. For a given differential scanning calorimetry curve with the heating rate β, one observes the maximum reaction rate at the peak temperature T.

In this study, the kinetic parameters of solid state transformations are determined from the maximum reaction rate at the peak temperature T. For a set of differential scanning calorimetry curves under a constant heating rate, the kinetic transformation is described with the mathematical expression of Kissinger Eq. 2 [27]:

$$\frac{d\left(\frac{\ln\beta}{T^2}\right)}{d\left(\frac{1}{T}\right)} = -\frac{E}{R} \tag{2}$$

where β is the heating rate, E_a is the activation energy, T is the absolute temperature corresponding to maximum process rate, and R is the gas constant.

For the processing the choice of dwell temperature was taking in account Tammann's temperature described as the temperature above which its constitutive cations become mobile so that their bulk diffusion is possible [28, 29].

The electrical conduction phenomenon in perovskite materials is very important since many properties depend on it. There is a strong correlation between the electrical conductivity of the materials, temperature, and the nature of the sample analyzed.

In bulk materials, two types of conductivity phenomena occur: long-range conductivity and localized transport oxygen vacancies. The conduction mechanism can be ionic, electronic, or both. The proportion of ionic to electronic conduction in the materials varies upon temperature and the purity of material. The variation of electrical conductivity with temperature is explained by Eq. (3):

$$\sigma = A\exp^{\frac{-E_a}{KT}} + B\exp^{\frac{-E_{ab}}{KT}} \tag{3}$$

Eq. (4) describes the phenomena at higher temperatures where the intrinsic conduction process dominates.

$$\sigma = A\exp^{\frac{-E_a}{KT}} \tag{4}$$

For both equations, E_a and E_b are the activation energy for the intrinsic and extrinsic conduction processes, respectively, A and B are constants, T is the absolute temperature, and K is the Boltzmann constant.

2. Experiment

2.1 Material preparations

The formation of the perovskite phase via solid state reaction was done under controlled rates for heating and cooling as well as for the dwell of thermal treatment. A 2°C/min rate of heating and cooling was selected. Preparatory trials were carried out in order to find the suitable temperature for dwell covering 600–1000°C.

Equimolar quantities of La_2O_3 and Co_3O_4 powders from Aldrich, with >99.8% purity, were mixed and ground thoroughly in agate mortar. The mixing was done with isopropanol from Chimopar SA, purity >96%. For the thermal treatment, the powder in alumina crucibles was placed in chamber furnace. Once the powder was synthesized, compacted ceramics were prepared by uniaxial pressing technique. After this, they were submitted for sintering. Similarly, preparatory trials were performed to find out the proper sintering thermal treatment.

2.2 Materials characterization

The crystalline structures of the prepared powders were characterized by heating in chamber furnace, environmental air atmosphere, at 50°C increments, making X-ray diffraction measurements. A MiniFlex 600 Rigaku analyzer was used. 2θ scans were recorded between 5 and 90° and a speed of 1°/min, with resolution of 0.1°/step. The as-obtained reflexion patterns are indexed by using the Inorganic Crystal Structure Database (ICSD). Lattice constants and quantitative values for the identified phases are obtained from the fit to the corresponding X-ray diffraction spectra by using the PDXL powder diffraction analysis package from Rigaku. The microstructure and morphology of the as-prepared powders were examined using a scanning electron microscope VP CARL ZEISS (Field Emission Scanning Electron Microscope—FESEM) with LaB_6 cathode enabling 2-nm resolution. Specimens of powders were prepared by depositing it on a conductive carbon-based double-faced adhesive tape. Differential scanning calorimetry (DSC) and thermogravimetric analysis (TGA) measurements were carried out on precursors and their equimolar mixture using STA 449 F5 Jupiter® from NETZSCH-Gerätebau GmbH. The instrument is equipped with Proteus® software to carry out the measurement and evaluation of the resulting data. The DSC/TG curves are recorded up to 1273 K, with a heating rate of 10 K/min.

The analyses of electrical properties of the sample were carried out from 133 to 513 K using Alpha-A Novocontrol Technologies Novocontrol GmbH analyzer in conjunction with a continuous nitrogen flow cryostat.

3. Results

3.1 Thermal behavior

Thermal analysis results of precursors, **Figure 2**, indicate that Co_3O_4 powder, in red color, is stable below 1023 K while La_2O_3 powder, in black color, exhibits several steps of mass loss during the heating regime. La_2O_3 precursor features three weight-loss regions. The first, a gradual weight loss not shown here, is related to the dehydration of free and physically absorbed molecular water and structural water from the precursor powder. The second and third regions, more pronounced, are associated with the dehydration process from $La(OH)_3$ to La_2O_3 via $LaOOH^-$, according to the chemical reactions described by Eqs. (5) and (6), respectively:

$$La(OH)_3 \rightarrow LaOOH^- + H_2O \tag{5}$$

$$2LaOOH^- \rightarrow La_2O_3 + H_2O \tag{6}$$

The total percent weight loss is around 1.88%. The first step, not shown here, of weight loss was 0.04%. The next steps correspond to the decomposition of $La(OH)_3$. The weight loss in following step was around 0.87%, and in the third step was 0.98%. However, it is well known that $La(OH)_3$ can react relatively fast with the water of the atmosphere to form hydroxyl phase [28]. The corresponding endothermic peaks on the differential scanning calorimetry curve appeared at 581 K and 910 K respectively. The corresponding areas are −23.44 and −15.34 μVs/mg, respectively.

The mixture of Co_3O_4 and La_2O_3 powder precursor, as can be seen in **Figure 3**, has a weight-loss step, associated with the presence of the La_2O_3, and after 1023 K, an endothermic event is detected around 1073 K. Previous studies have found similar features [30]. This is attributed to the ongoing dehydration of LaO(OH) and formation of $LaCoO_3$ from a solid state reaction of lanthanum and cobalt oxides, following Eq. (7):

$$La_2O_3 + Co_3O_4 \rightarrow LaCoO_3 \tag{7}$$

No significant thermic effect was registered at higher temperatures. The corresponding endothermic peaks on the differential scanning calorimetry curve appeared at 1073 K and the area of −147.7, −36.58, and −48.09 μVs/mg, respectively.

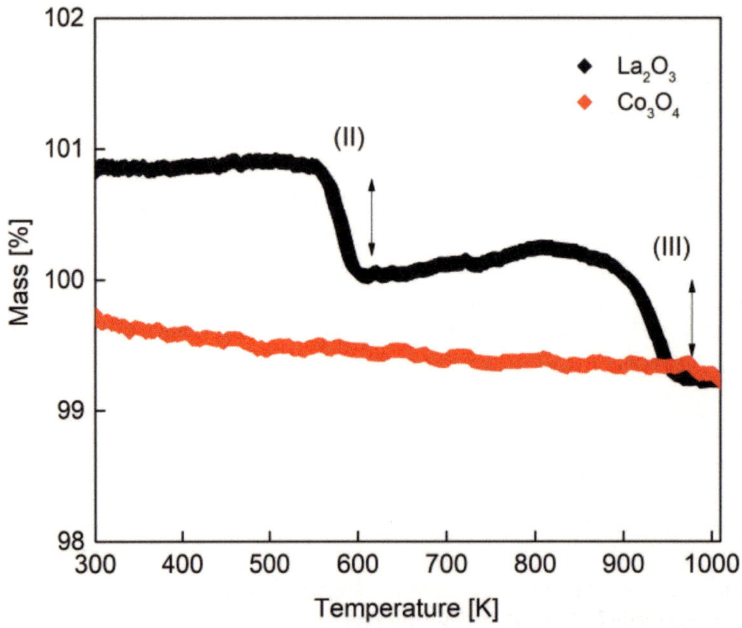

Figure 2.
Weight-loss % diagram for Co_3O_4 powder, in red color, and La_2O_3 powder, in black color.

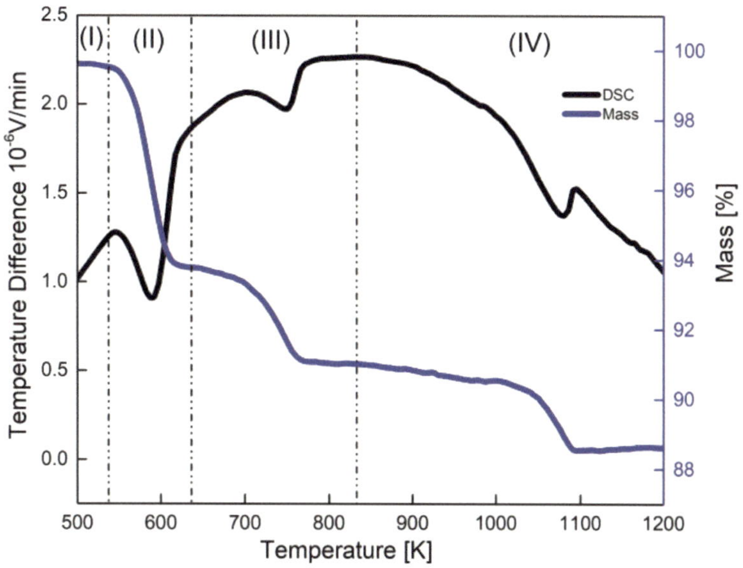

Figure 3.
DSC-TGA plot of Co_3O_4 and La_2O_3 precursor powders in mixture.

The kinetic analysis of phase change for a series of non-isothermal tests, **Figure 4**, is applied to estimate the activation energy. This analysis was performed for third observed exothermal reaction from differential scanning calorimetry data of phase transformation process. The peak temperature T depends on the evolution of the sample with the heating rate. The dynamic heating rate influences the observed peak temperature T of thermal events. These correspond to the variations of growth and phase change rate with temperature. The new phase formation rate depends on temperature and is independent of time. It is observed, in this reaction, the peak temperature has the tendency to slightly shift to the higher temperatures as the heating rates increase.

For polycrystalline reactants, the kinetic process depends on factors like particle size and shapes of the reacting particles, overlapping effects during the formations of the new phase, etc. Solid state reactions are complex and the evaluation of kinetic parameters is helpful for the development of approaching the preparation. In this evaluation, we made the assumption of first order of reaction. From the plot of rate of reaction $\ln(\beta/T^2)$ versus the reciprocal of absolute temperature, **Figure 5**, of as-mixed Co_3O_4 and La_2O_3 powder precursors, an E_a = 223 kJ/mol is obtained, which seems to be in good agreement with earlier results reported [31].

3.2 X-ray diffraction investigation

The analysis of X-ray diffraction pattern corresponding to unmixed and mixed precursors as function of temperature treatment is shown in **Figure 6**. In the lowest position of the diagram, the unmixed La_2O_3 and Co_3O_4 powder precursors are displayed. Going up in the diagram, the mixture of La_2O_3 and Co_3O_4 powder precursors at room temperature is indicated. The La_2O_3 powder precursor is tetragonal La_2O_3 (space group P-3m1) with a = b = 3.933 Å and c = 6.132 Å [32]. Traces of hexagonal $La(OH)_3$ (space group P63/m) are detected. The corresponding lattice constant values are a = b = 6.529 Å and c = 3.859 Å. Co_3O_4 powder indicates that the precursor is single-phase, with cubic crystal symmetry (space group Fd-3m) [32]. The lattice constant value obtained from fit to data is a = 8.038 Å.

Upon thermal treatment, which is done to favor the formation of end products of perovskite of $LaCoO_3$, the X-ray diffraction patterns show the formation of the perovskite phase. The XRD data confirm the perovskite $LaCoO_3$ phase is converted upon heat treatment starting at 1023 K, growing progressively with increasing temperature up to 1273 K. This is followed by a gradual smoothing of the peak intensities of the precursor phases. The resulting powder can be indexed on the basis of a rhombohedral unit cell; the peaks around $2\theta = 33.12°$ are bifurcated,

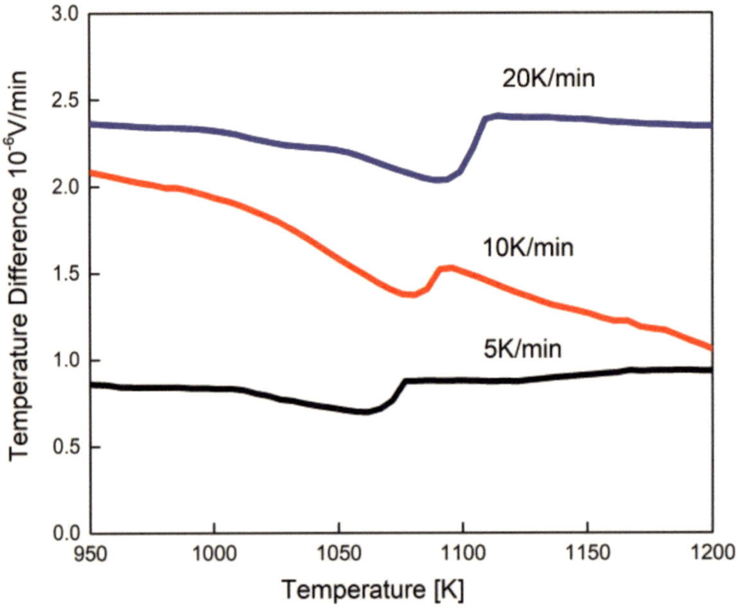

Figure 4.
Differential scanning calorimetry curves of as-mixed Co_3O_4 and La_2O_3 powder precursors at different heating schedules.

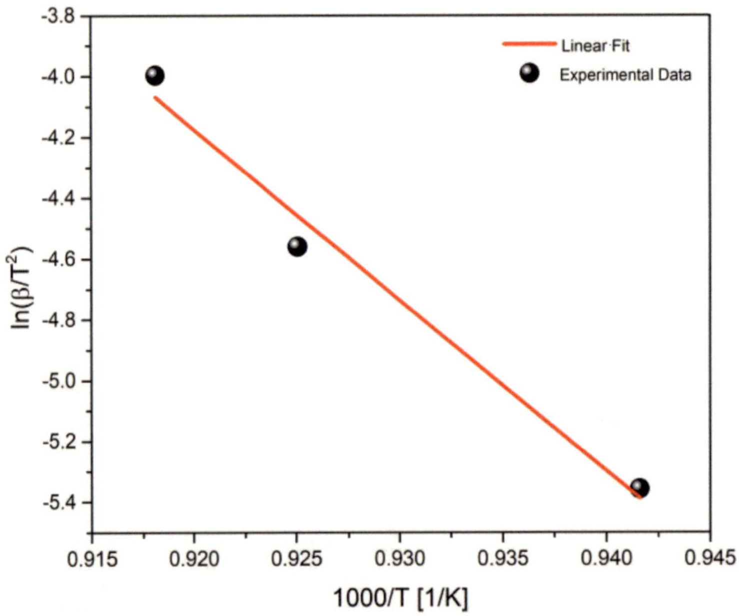

Figure 5.
Variation of $ln(\beta/T^2)$ plot for a set of differential scanning calorimetry scans with different heating rates.

which is the characteristic peak for the rhombohedral structure and confirmed by the experimental X-ray results as well as previous studies [33, 34].

The prepared $LaCoO_3$ has trigonal crystal symmetry space group R-c3 [35, 36]. The lattice constant values obtained from fit to data are a = b = 5.444 Å and c = 13.104 Å.

The average particle size is calculated from the XRD pattern using Debye and Scherrer formula, Eq. (8):

$$d = \frac{K\lambda}{\beta \cos\theta} \quad (8)$$

where K = 0.94 is the Scherrer constant, λ is the wavelength of the X-ray source used, β is the full width at the half maxima, and θ is the Bragg diffraction angle.

The result obtained for the average crystallite size by Scherrer's equation is 294 Å.

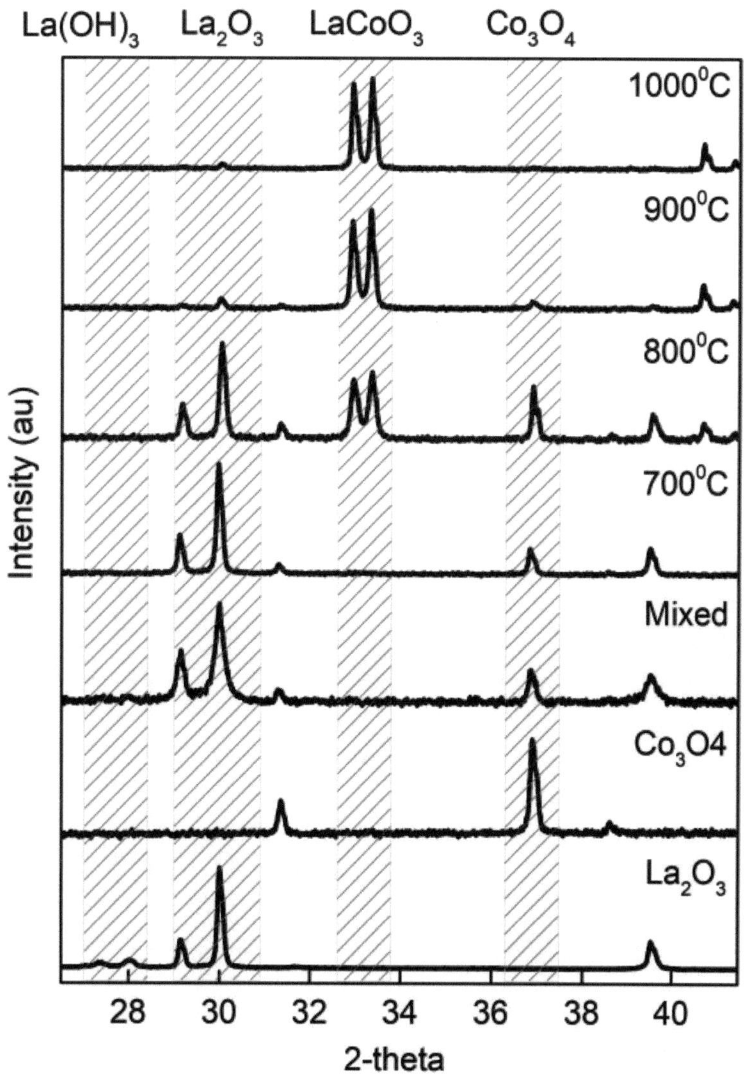

Figure 6.
Room temperature X-ray diffraction patterns of un/mixed La_2O_3 and Co_3O_4 powder precursors and $LaCoO_3$ powders according to the thermal treatment.

3.3 Microscopy analysis

Scanning electron micrograph of $LaCoO_3$ powders obtained by solid state synthesis at 1273 K is displayed in **Figure 7**. The perovskite $LaCoO_3$ powders obtained consist of pre-sintered conglomerates of grains. The agglomeration tendency is due to the high calcination temperature employed to obtain the single phase. The reaction product has well-defined shape and less uniform grain size distribution. The high-temperature calcination results in severe agglomeration. The sintering of many crystallites during high thermal treatment resulted in the larger grains, which are polycrystals. The grain limits and triple junction points are well defined.

3.4 Electric properties

In **Figure 8**, we show the electrical conductivity of our polycrystalline pellets pressed and sintered at 1173 K. Greater conductivity differences are noticed to depend on the temperature. The tendency of $LaCoO_3$ electrical conductivity is to increase when temperature is rising. This behavior of electrical conductivity in relation to the temperature indicates the process is thermally activated, which is characteristic for semiconductors. For our $LaCoO_3$ sample, semiconducting behavior appears at temperatures above 353 K, whereas metallic behavior was observed at temperatures below 353 K. This is in accordance with the previous statements that $LaCoO_3$ is ionic at elevated temperatures. The electrical conductivity is p-type for both ambient and elevated temperatures. With semiconductors, there are insufficient mobile carriers at low temperatures and resistance is high; but, as we heat the material, more and more of the lightly bound carriers escape and become free to conduct. Electrons are excited over the band gap and occupy energy levels in conductivity band whereas holes are created in valence band. The band gap in the metal is small and the electrons can easily jump to conduction band. For the metallic behavior with increasing temperature above absolute zero, the flowing electrons

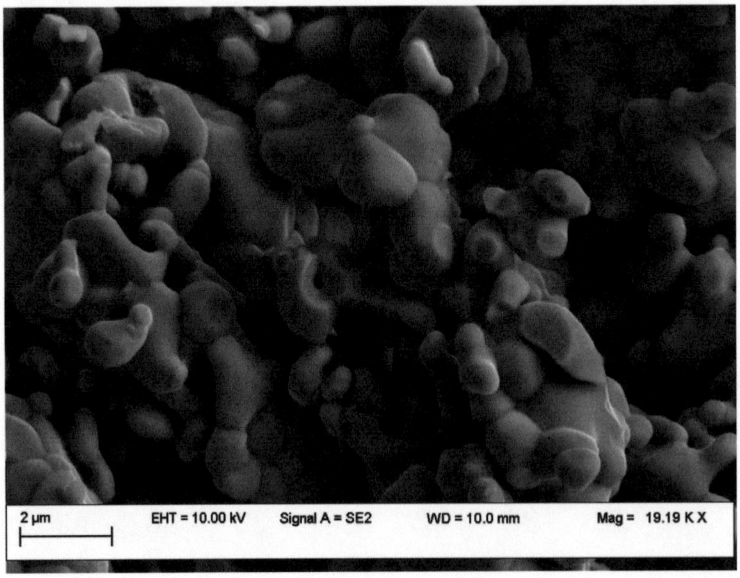

Figure 7.
SEM micrograph of $LaCoO_3$ powder.

will run into the atoms in the lattice. This will cause the atoms to move slightly out of their lattice sites and to interfere with electrons that travel freely. Effectively, they start to block electrons on their path, causing electrons to scatter. The change in slope of the conductivity-temperature plot is assumed to be the onset of ionic conductivity. $LaCoO_3$ is a mixed conductor with contributions from ionic conductivity and electronic conductivity.

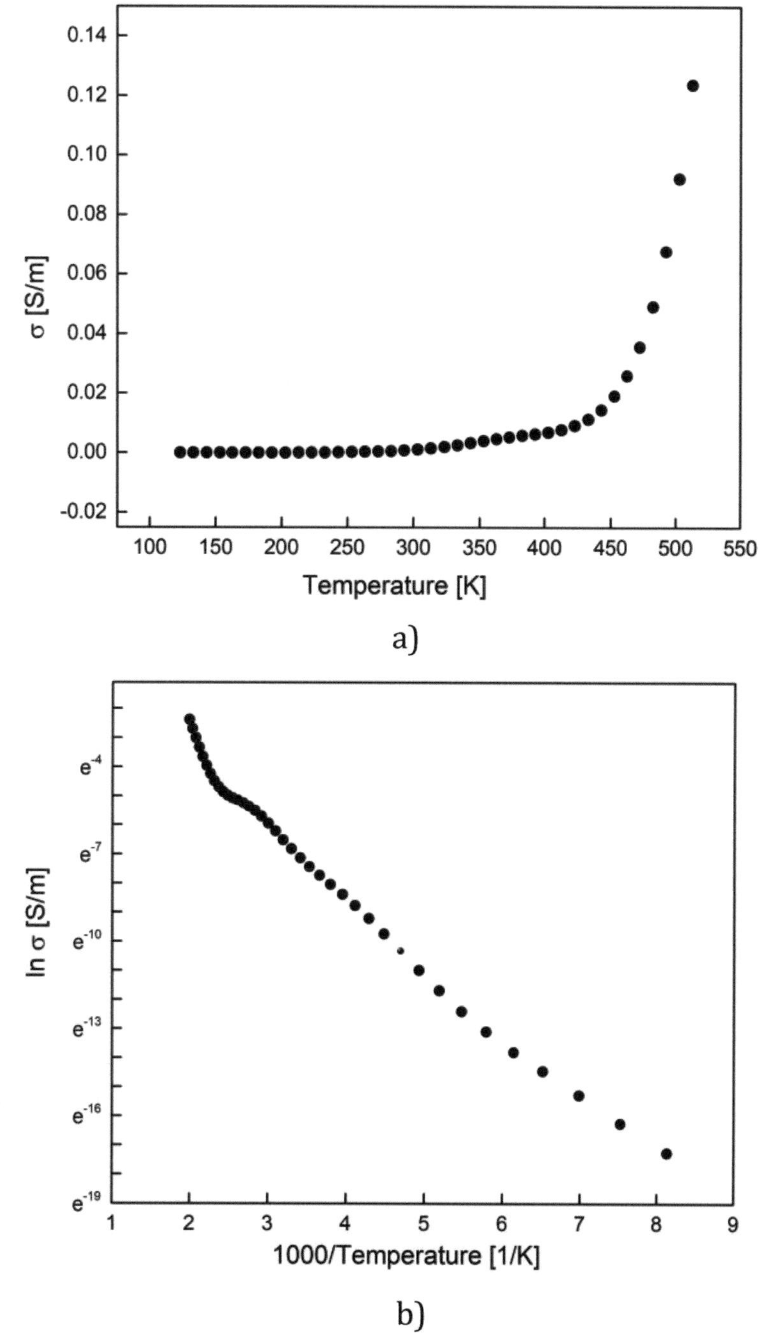

Figure 8.
$LaCoO_3$ total conductivity σ depending on absolute temperature in (a) and lnσ depending on absolute temperature in (b).

Variation of the electrical conductivity with temperature obeys Arrhenius formula, Eq. (9), and is calculated based on the following:

$$\sigma = \sigma_0 \exp\left(\frac{-E_a}{KT}\right) \tag{9}$$

where σ_0 is the pre-exponential factor, E_a is the activation energy, K is the Boltzmann constant, and T is the absolute temperature. The activation energy of the electrical conductivity was found to be E_a = 74 kJ/mol.

For the undoped $LaCoO_3$ oxide, the oxygen non-stoichiometry is considered to determine the carriers for the conduction phenomena. This is believed to be caused by the oxygen elimination with increase in the temperature (**Figure 9**).

The Kröger-Vink set of conventions are used to describe electric charge and lattice position for point defect species in crystals. The following ions and vacancies are used: La^{3+}, Co^{2+}, Co^{3+}, Co^{4+}, O^{2-}, oxygen vacancies. Making use of these in Eq. (10), the defect equilibrium between the charge species and oxygen partial pressure can explain the increase of partial pressure of oxygens:

$$2Co_{Co}^{\bullet} + O_O^x = V_O^{\bullet\bullet} + 2Co_{Co}^x + \frac{1}{2}O_2\,(g) \tag{10}$$

where $Co_{Co}^x = Co^{3+}$, $Co_{Co}^{\bullet} = Co^{4+}$ and $O_O^x = O^{2-}$.

Eq. (11) expresses the charge disproportion defect as follows:

$$2Co^{3+} \leftrightarrow Co^{2+} + Co^{4+} \tag{11}$$

When a voltage is applied to our sintered $LaCoO_3$ sample, a current flow is observed. At a given temperature, the type of charge carrier species is not defined. Because of this, it is useful to depict the conductivity dependence on temperature. The existence of ions is assured by definition, mixed conductors being ionic

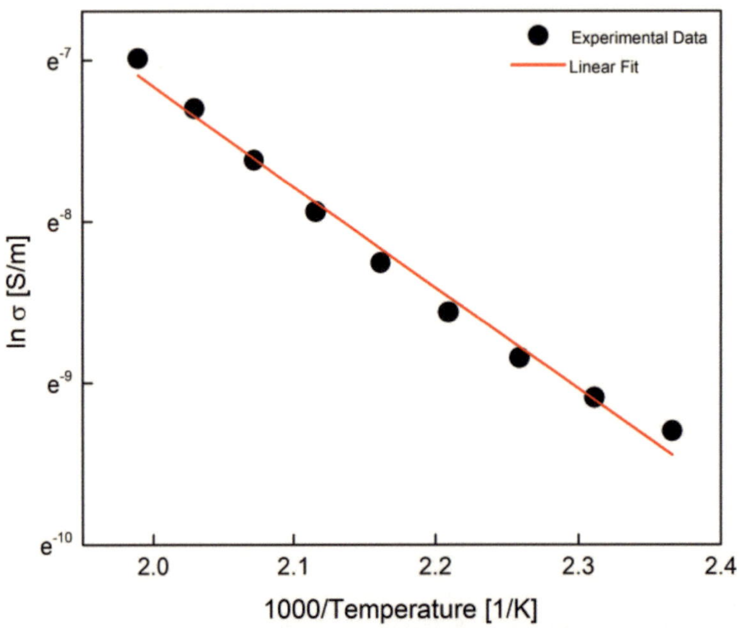

Figure 9.
Variation of dc conductivity (σ) with inverse of absolute temperature for $LaCoO_3$.

compounds. Ionic defects have their most pronounced effect on transport properties, which are ionic conductivity and diffusion. For $LaCoO_3$ samples, the oxygen vacancies are the mobile ionic defects.

4. Conclusions

In this study, we successfully used solid state synthesis as a dry method for preparation of rhombohedral lanthanum cobaltite $LaCoO_3$ perovskite-type oxides.

The solid state synthesis process is thermally activated and requires high temperatures. The X-ray diffraction analysis suggests that $LaCoO_3$ phase nucleates readily at 973 K. The formation of $LaCoO_3$ occurs gradually with increasing temperature, at the expense of precursors' consumption. At 1273 K, the reaction is essentially complete. Thermal analysis suggests that the reaction is accompanied by reversible transformation of $La(OH)_3$ into La_2O_3, which is a native surface phase of the oxide phase. The reaction kinetics is strongly mediated by Co_3O_4 transformation. Co_3O_4 starts to transform before its decomposition temperature. The overall electronic transport properties exhibited by the prepared powders are typically ionic at elevated temperatures (above 353 K) and are mediated by presence of oxygen vacancy.

The interrelationship between structure and properties is a key element of materials science and engineering. Thinking of the variety of oxide perovskite materials, each particular application is based on a certain functional property of the material. Most macroscopic properties observed in bulk (e.g., crystalline quality, ionic conductivity, density, etc.) are the consequence of the constituent particles' properties of the material. For applications, it is essential to know the relation between material microstructure and the macroscopic properties in order to adjust its performance to more specific requirements and also to keep the fabrication costs low. Properties like ionic conductivity, for example, make the perovskite material promising to be used as a solid electrolyte for solid oxide fuel cells (SOFC). Similarly, ferroelectric and piezoelectric behaviors lead to potential applications for thermistor, actuator, and piezoelectric transducers, and the property of superconductivity to superconductors.

We consider this method to be beneficial for both environmental protection and industrial applications because it does not have by-products or waste at the end of the reaction and the precursors are not expensive. In addition to this, an important advantage of this method is the sustainability to scale up the process. We recommend this for batch production which allows flexibility to adjust the process as required for a specific application. There is no need for additional facilities to collect the gaseous by-products or the waste to collect it and dispose it.

With appropriate optimization, this method we applied to $LaCoO_3$ can be applied to other perovskites; the synthesizing parameters like reaction time and temperature should be obtained by exploration as performed with the recipe presented in this chapter.

We conclude the present work stating that using this conventional method as solid state reaction allows obtaining highly single-phase $LaCoO_3$ powders for applications and uses of advanced functional materials.

Acknowledgements

This work was supported by the National Authority for Scientific Research and Innovation/Romanian Ministry of Education and Research, project "RESTORE"—117/16.09.2016 ID/Cod My SMIS: P_37_595/104958.

Conflict of interest

The authors state that there is no conflict of interest associated with this work.

Thanks

The authors would like to thank Dr. Amalia Soare for taking the SEM photograph.

Author details

Mirela Dragan[1*], Stanica Enache[1], Mihai Varlam[1] and Konstantin Petrov[1,2]

1 National R&D Institute for Cryogenic and Isotope Technologies, Ramnicu Valcea, Romania

2 Academician Evgeni Budevski Institute of Electrochemistry and Energy Systems, Sofia, Bulgaria

*Address all correspondence to: mirela.dragan@icsi.ro

References

[1] Lin Q, Yang X, Lin J, Guo Z, He Y. The structure and magnetic properties of magnesium-substituted $LaFeO_3$ perovskite negative electrode material by citrate sol-gel. International Journal of Hydrogen Energy. 2018;**43**(28):12720-12729. DOI: 10.1016/j.ijhydene.2018.03.156

[2] Goswami S, Bhattacharya D. Magnetic transition at ~150 K in nanoscale $BiFeO_3$. Journal of Alloys and Compounds. 2018;**738**:277-282. DOI: 10.1016/j.jallcom.2017.12.107

[3] Presto S, Kumar P, Varma S, Viviani M, Singh P. Electrical conductivity of NiMo-based double perovskites under SOFC anodic conditions. International Journal of Hydrogen Energy. 2018;**43**(9):4528-4533. DOI: 10.1016/j.ijhydene.2018.01.066

[4] Zhang F, Zhang X, Jiang G, Li N, Hao Z, Qu S. H_2S selective catalytic oxidation over Ce substituted $La_{1-x}Ce_xFeO_3$ perovskite oxides catalyst. Chemical Engineering Journal. 2018;**348**:831. DOI: 10.1016/j.cej.2018.05.050.p

[5] Wang WL, Meng Q, Xue Y, Weng X, Sun P, Wu Z. Lanthanide perovskite catalysts for oxidation of chloroaromatics: Secondary pollution and modifications. Journal of Catalysis. 2018;**366**:213-222. DOI: 10.1016/j.jcat.2018.07.022

[6] Fang M, Yao X, Li W, Li Y, Shui M, Shu J. The investigation of lithium doping perovskite oxide $LiMnO_3$ as possible LIB anode material. Ceramics International. 2018;**44**(7):8223-8231. DOI: 10.1016/j.ceramint.2018.02.002

[7] Yao C, Zhang H, Liu X, Meng J, Zhang X, Meng F, et al. Characterization of layered double perovskite $LaBa_{0.5}Sr_{0.25}Ca_{0.25}Co_2O_{5+\delta}$ as cathode material for intermediate-temperature solid oxide fuel cells. Journal of Solid State Chemistry. 2018;**265**:72-78. DOI: 10.1016/j.jssc.2018.05.028

[8] Heikes R, Miller RC, Mazelsky R. Magnetic and electrical anomalies in $LaCoO_3$. Physica. 1964;**30**(8):1600-1608. DOI: 10.1016/0031-8914(64)90182-X

[9] Menyuk N, Dwigh K, Raccah PM. Low temperature crystallographic and magnetic study of $LaCoO_3$. Journal of Physics and Chemistry of Solids. 1964;**28**(4):549-556. DOI: 10.1016/0022-3697(67)90085-6

[10] Enache S, Dragan M, Soare A, Ebrasu DI, Zaulet A, Varlam M, et al. One step solid-state synthesis of lanthanum cobalt oxide perovskites as catalysts for oxygen evolution in alkaline media. Special Issue of Bulgarian Chemical Communications, Sofia, Bulgaria. 2018;**50**(A):127-132

[11] Ebrasu DI, Zaulet A, Enache S, Dragan M, Carcadea E, Varlam M, et al. Electrochemical characterization of metal oxide as catalysts oxygen evolution in alkaline media. Special Issue of Bulgarian Chemical Communications, Sofia, Bulgaria. 2018;**50**(A):133-138

[12] Cullity BD, Graham CD. Introduction to Magnetic Materials. 2nd ed. Hoboken, New Jersey, United States: Wiley-IEEE Press; 2008. 171p. DOI: 10.1002/9780470386323

[13] Greenwood NN, Earnshaw A, editors. Chemistry of the Elements. 2nd ed. Oxford, United Kingdom: Butterworth-Heinemann; 1997. 1113p. DOI: 10.1016/B978-0-7506-3365-9.50032-8

[14] Zhu H, Zhang P, Dai S. Recent advances of lanthanum-based perovskite oxides for catalysis. ACS

Catalysis. 2015;**5**(11):6370-6385. DOI: 10.1021/acscatal.5b01667

[15] Chainani A, Mathew M, Sarma DD. Electron-spectroscopy study of the semiconductor-metal transition in $La_{1-x}Sr_xCoO_3$. Physical Review B. 1992;**46**(16):9976-9983. DOI: 10.1103/PhysRevB.46.9976

[16] Ma CL, Cang J. First principles investigation on the band gap of the ground state of $LaCoO_3$. Solid State Communications. 2010;**150**(41-42): 1983-1986. DOI: 10.1016/j.ssc.2010.08.023

[17] Shannon RD. Revised effective ionic radii and systematic studies of interatomic distances in halides and chalcogenides. Acta Crystallographica, Section A: Crystal Physics, Diffraction, Theoretical and General Crystallography. 1976;**32**(5):751-767. DOI: 10.1107/S0567739476001551

[18] Radaelli PG, Cheong SW. Structural phenomena associated with the spin-state transition in $LaCoO_3$. Physical Review B. 2002;**66**(9):094408-094417. DOI: 10.1103/PhysRevB.66.094408

[19] Kobayashi Y, Mitsunaga T, Fujinawa G, Arii T, Suetake M, Asai K, et al. Structural phase transition from rhombohedral to cubic in $LaCoO_3$. Journal of the Physical Society of Japan. 2000;**69**(1):3468-3469. DOI: 10.1143/JPSJ.69.3468

[20] Mizusaki J, Mima Y, Yamauchi S, Fueki K, Tagawa H. Nonstoichiometry of the perovskite-type oxides $La_{1-x}Sr_xCoO_{3-\delta}$. Journal of Solid State Chemistry. 1989;**80**(1):102-111. DOI: 10.1016/0022-4596(89)90036-4

[21] Xu R, Su Q. High-temperature synthesis. In: Xu R, Pang W, Huo Q, editors. Modern Inorganic Synthetic Chemistry. Amsterdam, Netherlands: Elsevier; 2011. pp. 9-38. DOI: 10.1016/B978-0-444-53599-3.10002-2

[22] Rojac T, Kosec M. Mechanochemical synthesis of complex ceramic oxides. In: Sopicka-Lizer M, editor. High-Energy Ball Milling. Sawston, Cambridge, United Kingdom: Woodhead Publishing; 2010. pp. 113-148. DOI: 10.1533/9781845699444.2.113

[23] Huízar-Félix AM, Hernández T, de la Parra S, Ibarra J, Kharisov B. Sol–gel based Pechini method synthesis and characterization of $Sm_{1-x}Ca_xFeO_3$ perovskite $0.1 \leq x \leq 0.5$. Powder Technology. 2012;**229**:290-293. DOI: 10.1016/j.powtec.2012.06.057

[24] Manukyan KV. Solution combustion synthesis of catalysts. In: Borovinskaya IP, Gromov AA, Levashov EA, Maksimov YM, Mukasyan AS, Rogachev AS, editors. Concise Encyclopedia of Self-Propagating High-Temperature Synthesis. Amsterdam, Netherlands: Elsevier; 2017. pp. 347-348. DOI: 10.1016/B978-0-12-804173-4.00137-X

[25] Sakka S. Sol–gel process and applications. In: Somiya S, editor. Handbook of Advanced Ceramics. 2nd ed. Cambridge, Massachusetts, United States: Academic Press; 2013. pp. 883-910. DOI: 10.1016/B978-0-12-385469-8.00048-4

[26] Agrawa D. Microwave sintering of ceramics, composites and metal powders. In: Fang ZZ, editor. Sintering of Advanced Materials. Sawston, Cambridge, United Kingdom: Woodhead Publishing Series in Metals and Surface Engineering; 2010. pp. 222-248. DOI: 10.1533/9781845699949.2.222

[27] Kissinger HE. Reaction kinetics in differential thermal analysis. Analytical Chemistry. 1957;**29**(11):1702-1706. DOI: 10.1021/ac60131a045

[28] Tammann G, Mansuri QA. Metallographische Mitteilungen aus dem Institut für physikalische Chemie der Universität Göttingen CXIII. Zur

Rekristallisation von Metallen und Salzeneitschrift für anorganische und allgemeine Chemie. 1923;**126**(1):119-128. DOI: 10.1002/zaac.19231260109

[29] Tammann G. Zur Molekular-Dynamik in Kristallen. Nachrichten von der Gesellschaft der Wissenschaften zu Göttingen, Mathematisch-Physikalische Klasse. Göttingen, Germany: Deutsche Forschungsgemeinschaft e.V; Vol. 281930. pp. 227-254. Available http://www.digizeitschriften.de/dms/resolveppn/?PID=GDZPPN002508184

[30] Jiawen D, Yanli W, Weili S, Youngxiu L. Preparation of $La(OH)_3$ and La_2O_3 with rod morphology by simple hydration of La_2O_3. Journal of Rare Earths. 2006;**24**(4):440-442. DOI: 10.1016/S1002-0721(06):60139-7

[31] Walter D. Kinetic analysis of the transformation from lanthanum hydroxide to lanthanum oxide. Zeitschrift für Anorganische Chemie. 2006;**632**(12-13):2165-2165. DOI: 10.1002/zaac.200670177

[32] Aldebert P, Traverse JP. Etude par diffraction neutronique des structures de haute temperature de La_2O_3 et Nd_2O_3. Materials Research Bulletin. 1979;**14**(3):303-323. DOI: 10.1016/0025-5408(79)90095-3

[33] Liu X, Prewitt CT. High-temperature X-ray diffraction study of Co_3O_4: Transition from normal to disordered spinel. Physics and Chemistry of Minerals. 1990;**17**(2):168-172. DOI: 10.1007/BF00199669

[34] Ozawa M, Onoe R, Kato H. Formation and decomposition of some rare earth (RE = La, Ce, Pr) hydroxides and oxides by homogeneous precipitation. Journal of Alloys and Compounds. 2006;**408-412**:556-559. DOI: 10.1016/j.jallcom.2004.12.073

[35] Dragan M, Misture S. In-situ analysis of chemical expansion and stability of SOFC cathodes. Materials Research Society Symposium Proceedings. 2014;**1655**:77-82. DOI: 10.1557/opl.2014.413

[36] Haas O, Struis RPWJ, McBreen JM. Synchrotron X-ray absorption of $LaCoO_3$ perovskite. Journal of Solid State Chemistry. 2004;**177**(3):1000-1010. DOI: 10.1016/j.jssc.2003.10.004